Vue.js 3 for Beginners

Learn the essentials of Vue.js 3 and its ecosystem to build modern web applications

Simone Cuomo

‹packt›

Vue.js 3 for Beginners

Group Product Manager: Kaustubh Manglurkar

Publishing Product Manager: Vaideeshwari Roshan

Senior Content Development Editor: Feza Shaikh

Technical Editor: Reenish Kulshrestha

Copy Editor: Safis Editing

Project Coordinator: Aishwarya Mohan

Proofreader: Safis Editing

Indexer: Subalakshmi Govindhan

Production Designer: Alishon Mendonca

Marketing Coordinators: Anamika Singh and Nivedita Pandey

First published: August 2024

Production reference: 3010925

Published by Packt Publishing Ltd.
Grosvenor House
11 St Paul's Square
Birmingham
B3 1RB, UK

ISBN 978-1-80512-677-5

www.packtpub.com

To my amazing wife, who gives me the drive to always reach for the stars even if this means having to go to sleep with someone typing noisily next to her, and to my kids, who are my biggest fans!

– Simone Cuomo

Foreword

I have had the privilege of working closely with Simone Cuomo for many years. He is currently the VP of software delivery at This Dot. Simone's exceptional skills as a mentor, architect, and executive have made him a standout figure in the software development community. His deep expertise in Vue.js has been invaluable to clients such as Wikimedia, guiding them through complex Vue.js migrations and new projects from the ground up.

Simone's passion for teaching and mentorship is evident in everything he does. He is committed to making coding and learning more accessible to everyone. In this comprehensive guide, *Vue.js 3 for Beginners*, Simone shares his vast knowledge and practical experience to help you become proficient in Vue.js.

Throughout the book, Simone offers clear, step-by-step instructions, complete with screenshots, source code snippets, and real-world examples. You'll discover how to build robust, maintainable applications using Vue.js, and gain insights into best practices for writing high-quality code.

Vue.js 3 for Beginners is more than just a technical manual; it's a gateway to becoming a better developer under the guidance of one of the industry's best mentors. Simone's dedication to mentorship and his engaging teaching style make this book an essential resource for anyone looking to excel in the world of Vue.js and beyond. With this guide, you'll be well-equipped to tackle any project and elevate your development skills to new heights.

Tracy Lee

CEO, This Dot

Contributors

About the author

Simone Cuomo is a VP of Delivery with a pure passion for programming. Simone works with various technologies but greatly excels in working with frontend tools, specifically with Vue.js, testing, PWA, performance, and accessibility.

Simone has been working with Vue.js for over seven years and has been an active member of the framework community, hosting remote events, providing corporate training, delivering conference talks, and writing about Vue.js in his blog, and for magazines and guest posts.

Simone has helped many companies improve their Vue.js codebases to make them scalable and performant. He prides himself on supporting the Wikimedia Foundation in adapting Vue.js and recently helped them migrate to Vue.js 3.

About the reviewers

Michael Di Prisco is an Italian developer and international speaker with over a decade of experience in the software world. Currently acting as a tech lead at Jointly, he started working in Ireland and then moved back to Italy to pursue a career as a full stack developer, before specializing in back-end development and software architecture.

Passionate about sharing knowledge, he actively makes many contributions to the open source community. He also enjoys writing articles on his website and collaborates with prominent Italian blogs.

In 2023, he embarked on a project called "Il Libro Open Source", with the purpose of writing an Open Source book about the software development world, helped by dozens of other contributors.

Adnan Kreiker is a frontend developer based in Berlin, Germany, with a passion for JavaScript and its ecosystem. He specializes in Vue.js, having been drawn to its beginner-friendly approach, powerful capabilities, and vibrant community. With over three years of experience, Adnan has worked on numerous Vue projects, including a large-scale SaaS application that helps reduce energy consumption and CO_2 emissions in both commercial and residential buildings. His expertise also includes migrating applications from Vue 2 to Vue 3.

Adnan is an avid supporter of the Vue.js and TypeScript communities. He has attended Vue Amsterdam for two consecutive years and presented at the 2023 TypeScript Global Summit, sharing his insights on Vue 3 and TypeScript integration. When reviewing technical books on front-end technologies, he seeks clear explanations, practical code examples, and a strong focus on real-world project applications. Adnan is a lifelong learner whose interests outside of coding include fitness, reading, and exploring new topics.

Table of Contents

Part 1: Getting Started with Vue.js

1

Exploring the Book's Layout and Companion App   **3**

2

The Foundation of Vue.js   **15**

Part 2: Understanding the Core Features of Vue.js

6

Event and Data Handling in Vue.js   93

7

Handling API Data and Managing Async Components with Vue.js   111

Part 3: Expanding Your Knowledge with Vue.js and Its Core Libraries

8

Testing Your App with Vitest and Cypress   133

9

Introduction to Advanced Vue.js Techniques – Slots, Lifecycle, and Template Refs   157

10

Handling Routing with Vue Router   177

Part 4: Conclusion and Further Resources

14

Advanced Resources for Future Reading   259

Index   271

Other Books You May Enjoy   280

Preface

The last decade has seen a surge of JavaScript frameworks aimed at improving client experiences with better **User Experience (UX)** and enhanced client interactivity. One of the most mature and popular frameworks is Vue.js.

Vue.js is a JavaScript framework for building user interfaces. It builds on top of standard HTML, CSS, and JavaScript and provides a declarative, component-based programming model that helps you efficiently develop user interfaces of any complexity.

Vue.js has become a favorite among developers of all levels for its vibrant ecosystem, delightful development experience, and focus on simplicity.

This book, *Vue.js 3 for Beginners*, is your comprehensive guide to mastering this framework. The aim of this book is to take you on a journey from the core concepts of Vue.js to building a real-world application, step by step.

There are many tutorials, articles, and documentation out there that share Vue.js features, but here is what sets this book apart:

- **Learning by doing**: Throughout the book, you'll be building a Companion App alongside the theoretical concepts. This hands-on approach ensures you grasp the connections between different sections and how Vue.js components work together.

- **Beginner-friendly**: Whether you're new to JavaScript frameworks or have some experience with others, this book provides a solid foundation in Vue.js. We start with the basics and gradually introduce more advanced techniques as you progress.

- **Real-world examples**: We'll use clear and practical examples to illustrate key concepts, making Vue.js more accessible and engaging. In many cases, multiple methods will be covered with their individual advantages and disadvantages. The book also includes many callouts to help you gain further context.

- **Written from experience**: This book is the result of many successful projects and numerous trained mentees. The flow in which the topics are introduced and explained has gone through many iterations across the span of my career.

By the end of this book, you'll not only have a strong understanding of Vue.js but also a fully functional application you can use or adapt for your own projects.

Who this book is for

Aspiring web developers, students, hobbyists, and anyone who wants to learn Vue.js from scratch and is eager to delve into front-end development using a modern and popular framework will benefit from this book. The book requires a basic understanding of front-end technologies such as HTML, CSS, and JavaScript. The topics are introduced in a modular fashion by breaking down problems into small, easy-to-understand units.

The book covers many topics of the Vue.js framework and defines good standards that can also be beneficial for existing Vue.js developers that want to ensure they use the framework correctly.

What this book covers

Chapter 1, Exploring the Book's Layout and Companion App, covers the details of the Companion App and what it will include. This chapter defines the structure and methodology used throughout the book and includes important topics not related to the framework such as component-based architecture, atomic design, and core web development areas.

Chapter 2, The Foundation of Vue.js, focuses on providing vital information about the Vue.js framework. The chapter covers topics such as the reactivity system, component composition, and framework lifecycles.

Chapter 3, Making Our HTML Dynamic, is where we begin the work required to build our Companion App. We will learn how to initialize a Vue.js application and see the basic steps required to turn a static HTML file into a dynamic Vue.js component.

Chapter 4, Utilizing Vue's Built-In Directives for Effortless Development, introduces one of the most important features of Vue.js: built-in directives. In this chapter, we will learn how to enhance the interactivity and dynamicity of our component by introducing the most important directives.

Chapter 5, Leveraging Computed Properties and Methods in Vue.js, teaches how to make our components concise and readable by introducing different techniques to handle data and logic within our components. This chapter introduces topics including Ref and Reactive data computed properties and methods that are the pillars of Vue.js logic.

Chapter 6, Event and Data Handling in Vue.js, expands our knowledge of the Vue.js framework by teaching us how to handle communication between components. We will also deepen our knowledge of props and introduce the notion of custom events.

Chapter 7, Handling API Data and Managing Async Components with Vue.js, teaches us how to deal with external asynchronous data. We will put into practice the notion of lifecycles and learn how to handle asynchronous data with `watch` and `<suspense>`.

Chapter 8, Testing Your App with Vitest and Cypress, will provide us with the tools necessary to ensure our code is well written by introducing testing. This chapter covers testing in general and then defines both unit tests using Vitest and E2E tests with Playwright for testing our existing Companion App components.

Chapter 9, Introduction to Advanced Vue.js Techniques – Slots, Lifecycle, and Template Refs, brings you back to the core of Vue.js and introduces advanced techniques. The chapters cover advanced topics such as slots, lifecycles, and Template Refs.

Chapter 10, Handling Routing with Vue Router, introduces the first external package: vue-router. In this chapter, we will add routing to our Companion App. We will learn how to define our router and how to navigate within our application.

Chapter 11, Managing Our Application's State with Pinia, focuses on state management and introduces a new core package: Pinia. We will learn how state management can simplify our application and create multiple example stores within our Companion App to learn about the different features that Pinia has to offer.

Chapter 12, Achieving Client-Side Validation with VeeValidate, introduces the topics of form handling and validation. We will review the Vue.js native tools for handling forms such as v-model, and cover advanced cases by introducing VeeValidate.

Chapter 13, Unveiling Application Issues with the Vue Devtools, takes a step back from the development of our Companion App and focuses on the debugging techniques required to improve our skills and understanding of the Vue.js framework.

Chapter 14, Advanced Resources for Future Reading, concludes our journey with some further reading and materials that will help you continue your learning of the Vue.js framework.

To get the most out of this book

The books cover in detail everything related to Vue.js and any additional package that is introduced, but it requires the reader to have a basic understanding of front-end development. Existing knowledge of HTML, CSS and JavaScript is required to make the most out of this book.

Software/hardware covered in the book	Operating system requirements
Vue.js 3	Windows, macOS, or Linux
Vite	Windows, macOS, or Linux
Crypress	Windows, macOS, or Linux
Vitest	Windows, macOS, or Linux
Vue DevTools	Windows, macOS, or Linux
Vue-router	Windows, macOS, or Linux
Pinia	Windows, macOS, or Linux
VeeValidate	Windows, macOS, or Linux

The book includes all the information required to install and configure the software used during the course of our journey. There are some dependencies that are not introduced in the book but required. The first is Node.js. The application will work with any major version of Node, so you can install the latest stable. The next requirement is an IDE. I will be using Visual Studio Code, but you can use any other IDE.

If you are using the digital version of this book, we advise you to type the code yourself or access the code from the book's GitHub repository (a link is available in the next section). Doing so will help you avoid any potential errors related to the copying and pasting of code.

The final chapter of the book provides you with guidance on how to continue your journey. I suggest you continue your learning journey to continue to improve your skills in Vue.js.

Download the example code files

You can download the example code files for this book from GitHub at `https://github.com/PacktPublishing/Vue.js-3-for-Beginners`. If there's an update to the code, it will be updated in the GitHub repository.

We also have other code bundles from our rich catalog of books and videos available at `https://github.com/PacktPublishing/`. Check them out!

Conventions used

There are a number of text conventions used throughout this book.

`Code in text`: Indicates code words in text, database table names, folder names, filenames, file extensions, pathnames, dummy URLs, user input, and Twitter handles. Here is an example: Vue.js tries to render the app with the data it has, resulting in the missing data being set as `null`."

A block of code is set as follows:

```
<template v-if="comments.length === 0"></template>
<template v-else></template>
```

When we wish to draw your attention to a particular part of a code block, the relevant lines or items are set in bold:

```
const posts = reactive([]);
const fetchPosts = () => {
   ...
}
fetchPosts();
```

Any command-line input or output is written as follows:

```
npm install
```

Bold: Indicates a new term, an important word, or words that you see onscreen. For instance, words in menus or dialog boxes appear in **bold**. Here is an example: " Because **About Page** does not require the overriding of the footer, we are leaving that out of our instance so that the default value can be rendered."

> **Tips or important notes**
> Appear like this.

Get in touch

Feedback from our readers is always welcome.

General feedback: If you have questions about any aspect of this book, email us at `customercare@ packtpub.com` and mention the book title in the subject of your message.

Errata: Although we have taken every care to ensure the accuracy of our content, mistakes do happen. If you have found a mistake in this book, we would be grateful if you would report this to us. Please visit `www.packtpub.com/support/errata` and fill in the form.

Piracy: If you come across any illegal copies of our works in any form on the internet, we would be grateful if you would provide us with the location address or website name. Please contact us at `copyright@packt.com` with a link to the material.

If you are interested in becoming an author: If there is a topic that you have expertise in and you are interested in either writing or contributing to a book, please visit `authors.packtpub.com`.

Share Your Thoughts

Once you've read *Vue.js 3 for Beginners*, we'd love to hear your thoughts! Scan the QR code below to go straight to the Amazon review page for this book and share your feedback.

`https://packt.link/r/1805126776`

Your review is important to us and the tech community and will help us make sure we're delivering excellent quality content.

Download a free PDF copy of this book

Thanks for purchasing this book!

Do you like to read on the go but are unable to carry your print books everywhere?

Is your eBook purchase not compatible with the device of your choice?

Don't worry, now with every Packt book you get a DRM-free PDF version of that book at no cost.

Read anywhere, any place, on any device. Search, copy, and paste code from your favorite technical books directly into your application.

The perks don't stop there, you can get exclusive access to discounts, newsletters, and great free content in your inbox daily

Follow these simple steps to get the benefits:

1. Scan the QR code or visit the link below

https://packt.link/free-ebook/9781805126775

2. Submit your proof of purchase
3. That's it! We'll send your free PDF and other benefits to your email directly

Part 1:
Getting Started
with Vue.js

In the first part of this book, we are going to introduce essential knowledge that will be required to complete your learning journey and gain a better understanding of the Vue.js framework. This book will set the foundation for the architecture used to build the Companion App and provide essential knowledge about Vue.js and its core logic.

This part contains the following chapters:

- *Chapter 1, Exploring the Book's Layout and Companion App*
- *Chapter 2, The Foundation of Vue.js*

1

Exploring the Book's Layout and Companion App

Vue.js is an enormously popular framework in the **JavaScript (JS)** ecosystem. In recent years, it has gained lots of popularity thanks to its simplicity, its great documentation, and, finally, its fantastic community. If you are starting web development now, or are transitioning from a different framework or language, Vue.js is a great choice.

Before we can jump into the main content of the book, it is important to learn how the book is structured and what methods will be used to explain the different topics of this fantastic framework.

To simplify the learning of Vue.js and make the book more interesting and interactive, the book has been built around the creation and enhancement of a Companion App.

Vue.js 3 for Beginners is going to focus mainly on the framework and its core libraries, and it will not cover basic development knowledge such as HTML, CSS, JS, and Git. To understand the content of this book, basic knowledge of these four topics is required.

The first part of this book is going to cover an important aspect of our learning journey and will provide you with important theoretical information that is needed for you to make the most of the book's content; we will then jump into the specifics of Vue by introducing the framework and its core concepts in *Chapter 2*. Finally, from *Chapter 3* onward, we will start to work on our application, one component at a time.

In this chapter, we'll cover the following topics:

- The Companion App
- The core areas of a web application
- Component-based architecture

By the end of this chapter, you will learn about what we are going to build during the course of the book, and cover some theoretical aspects required for us to make the most of the Vue.js framework, such as component-based architecture and the architectural decisions behind the Companion App.

Technical requirements

The application that accompanies the book has been built using free software and APIs and will not require you to purchase anything. However, there are some specific technical requirements needed for you to follow along:

- Visual Studio Code or another equivalent IDE (integrated development Environment)
- Volar Visual Studio code extension
- A browser updated to the latest version (I suggest Chrome or Firefox)
- Node 16+
- GIT or a Git **GUI (Graphic user interface)** such as GitKraken installed on your machine

The companion app

Learning a new language or framework is not an easy task. There are plenty of free resources, such as documentation, blogs, and YouTube videos on the internet, but I believe learning a new tech requires practice, and there is no better way to achieve this than by building a production-ready, performant, scalable social media application together.

The application is going to be very similar to the social media platform X (formerly Twitter). We will start from a clean canvas and slowly add more features and functionality until the app is fully working and ready to be added to your portfolio and showcased at your next job interview.

Each chapter will have a set of sections that will help you navigate the book. This will not only ensure that you can always follow along and have a clear understanding of the scope of the chapter, but it also allows you to use the book as a reference after you have read it all and allows you to jump to a specific chapter if you need to do so.

Each chapter includes the following sections:

- Starting branch for the chapter
- The current state of the Companion App
- Definition of what will be added and achieved within the current chapter
- Multiple sections of explanation and coding
- Summary of what Vue.js topics we have learned in this chapter with a glossary

The Companion App features

As mentioned previously, the application that we are about to build will be very similar to an existing social media application. To make sure we cover most of the Vue.js features and its ecosystem, we may at times over-architect a specific component or feature, but when this happens, it will be called out so that you will have full knowledge of whether it is a good practice to follow in the future and what would be the correct implementation.

By following the book, you will learn the following:

- How to structure a web application using a component-based architecture
- How to create simple and complex HTML using Vue.js
- How to make the right decision to make your app performant and scalable
- How to communicate between components
- How to use external APIs to load dynamic data
- How to use state management using Pinia
- How to implement multiple pages (routing) using vue-router
- How to test your application using Vitest and Cypress
- How to create forms effectively using Vue.js
- How to debug your application using the Vue debugger

The preceding list is just an overview of what we will be achieving in the book, and we are going to make this learning fun and interactive by building using the Companion App together.

The application code

The code for the application can be found in this repository: `https://github.com/PacktPublishing/Vue.js-3-for-Beginners`. If you do not know what a repository is or how to use it, I suggest you learn the basics, even if all the information and commands you require to use the code will be provided in each chapter.

The repository has multiple branches for each chapter. This will be the starting point for each chapter and will be specified at the start of each chapter, as mentioned before.

The main branch of the repository is the latest commit, and it includes the complete application. If you have some time, I suggest you run the full application to try and browse it to see what we will achieve in the course of the book.

To run the application, you can simply follow the instructions available in the `README.md` file that is available at the root of the project.

As it is the first time that we are running the application, I will also provide the information required here to get the application up and running:

1. First, we need to get a copy of the remote repository on our machine. To do so, run the following command in the terminal:

```
git clone https://github.com/PacktPublishing/Vue.js-3-for-Beginners
```

2. Then, we need to navigate into our newly created project folder:

```
cd vue-for-beginners
```

3. Before we can run the project, we need to install all its dependencies using a package manager. A package manager is a piece of software that is used to install and manage the packages, in our case Node.js and JS, for which the project depends. The application shared in the repository supports all major package managers, such as npm, yarn, and pnpm. In the following example, we are going to use npm:

```
npm install
```

4. Finally, it is time to run the project. The following command will run a development version of the project:

```
npm run serve
```

After a few seconds, the local instance of the application will start, and you should be able to access it by opening the browser at HTTP://localhost:5173. The application should look like this.

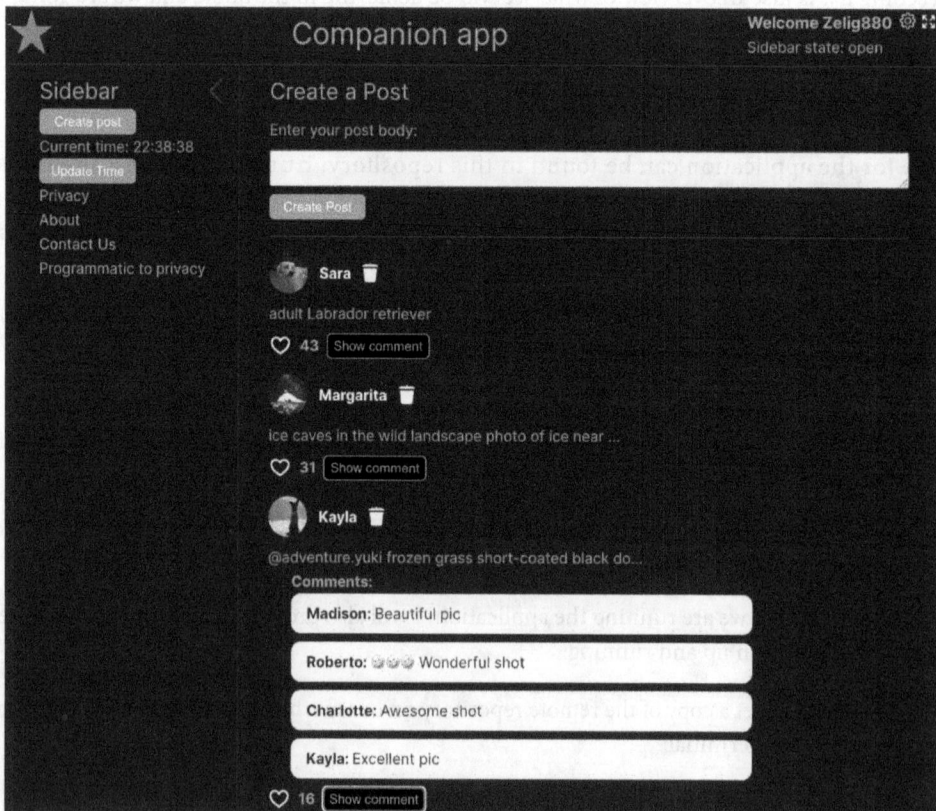

Figure 1.1: Screenshot of the Companion App dashboard

Spend some time navigating the application, both in the browser and within the code base, to see what we will build in the course of the book.

In this section, we have learned about the Companion Application, how it is going to be used to support our learning, its core features, and finally, the commands required to run the application locally. In the next section, we will spend a few moments on the core areas of web applications and explain which technologies/libraries we are going to use in our application.

The core areas of a web application

The JS ecosystem is not shy of frameworks and libraries, but even with this extensive choice, they mostly share the same core values and areas. These are the core parts of a web application and no matter which framework you use to write your application, you will have to know this and have a basic understanding of what they mean.

The pillars of a web application are as follows:

- **User interface (UI)**: This refers to the elements displayed on the screen from which a user can interact. In simple words, anything you can see or interact with on the internet is part of the UI. This core area of web development is usually achieved with basic HTML/CSS, vanilla JS (that is a different way to say plain JS), or a framework such as React, Vue, or Angular. In our case, this will be achieved using Vue.js 3.

- **Data fetching**: Data fetching is at the core of any web application. No matter how small your site is, it is going to require you to fetch some data. This technique is called data fetching and can either be achieved using a REST API or GraphQL. Data handling is not supported by any of the official libraries offered by Vue.js and we are going to handle it using plain JS with the `fetch` method.

- **State management**: Unless your website is a static blog post, you will need to handle some data. This could be the current state of a form or information about a logged-in user. Small applications can easily achieve this directly with the existing tool that a framework provides, but at times, this needs to be expanded to use full-blown "state management." In Vue.js, two main libraries help you handle your data, Vuex the state management of choice for Vue 2 and Pinia, which is the suggested library to be used for Vue 3 (Pinia is just a newer version of Vuex, but it was renamed due to the fact that it went through a full rewrite with many breaking changes). Because we will be writing our Companion App in Vue 3, we will use Pinia.

- **Routing**: Even if by definition most of today's websites are called **single-page applications** (**SPAs**), in reality, they make use of more than one page. The definition of "single page" is just used because the application does not fully reload during navigation, but it does not imply that the application will not have more than one route. For this reason, most web applications will require a way to handle routing between multiple pages. For the scope of this book, we will be using vue-router, which is the official routing library.

- **Forms and validation**: Forms are probably one of the main reasons why JS frameworks and SPAs have become so successful. The possibility of handling complex forms and client interaction without the need to refresh the page has improved **user experience** (**UX**) massively. Even if Vue.js is more than capable of handling forms and their validation, we will be using an external library called VeeValidate for client-side validation.

- **Debugging**: Building is not always straightforward and debugging an application is a must-have skill. Even if this is not really a real part of the application (as it is more a skill than an actual part of the application itself), I still want to include it as part of a web application core area, as debugging helps us make the application secure and performant. In our case, we will be using plain JS techniques and a browser extension called Vue.js devtools to help us analyze, study, and improve our application.

In this section, we explained the different areas that make a web application. We also explained the architectural decision behind the technology stack that is going to be used within our Companion App. It is now time to learn about a fundamental methodology called **component-based architecture**. This is the foundation for most frontend frameworks.

Component-based architecture

We have reached the final section of our introductory discussion and we are almost ready to start coding. This section is going to introduce the concept of component-based architecture. Even if you are already familiar with this topic, I suggest you continue reading this chapter as it will support some of the decisions we will make later in the course of the book.

In this section, we are going to cover how web development worked before this concept was introduced and we will then discuss how component-based architecture has shaped the web development industry as we know it.

One page at a time

If you have been a developer for as long as I have been, you have probably worked with languages and frameworks that were not flexible in the way the pages were defined and developed. Using .NET and PHP a few years ago would have meant that each web page was created using a single file (disclaimer: some languages had the definition of "partials".

This worked well until JS started to be used in the frontend and shook the ecosystem. JS changed sites from static pages to very dynamic entities and in doing so pushed for something more dynamic that would not work with the previous development tools.

Let's take into consideration a standard website homepage, such as the following one:

Figure 1.2: Wireframe of a standard homepage

This site follows a standard layout with a header and a footer, a banner, some featured content, and a **Contact Us** form. Imagine having all this in one single HTML file. Back in the day, this was probably all held in one single HTML file with a shared stylesheet, for example a CSS (Cascading Style Sheet) file. As mentioned previously, things started to change in the industry and JS started to be used more and more.

In the preceding scenario, JS was probably just used to add some basic interactivity like slideshow animation in the banner, some fancy scrolling i products list or to handle form submission in the **Contact Us** form.

Long story short, this change slowly shaped the industry toward frontend libraries and frameworks. These libraries and frameworks aimed to help manage and simplify the hundreds of lines of code produced in JS and they did so by introducing component-based architecture.

Breaking things down into small units was not something new in the industry, as the backend framework already had this notion with the use of **object-oriented programming**, but it was an innovation in the frontend side of the industry.

From one page to many components

The term **component-based development (CBD)** is a pattern in which the UI of a given application is broken down into multiple "components." Breaking down big pages into small individual units reduces the complexity of the application and helps the developer focus on the individual scope and responsibility of each section.

All of today's frontend frameworks are built on top of this pattern and today's frontend development is driven by architecture based on CBD.

Let's look at the previous example of the home page and see how we could split this into small isolated components.

The home page would be broken down into the following components:

- **Header**: A component that will include the logo and the logic used to display account information such as the avatar
- **The slideshow**: A reusable component used to display slideshow images
- **Featured**: A component used to display featured articles
- **Contact Us**: A component including all the logic required to validate and submit our form
- **Footer**: A static component that will include some copy and social links

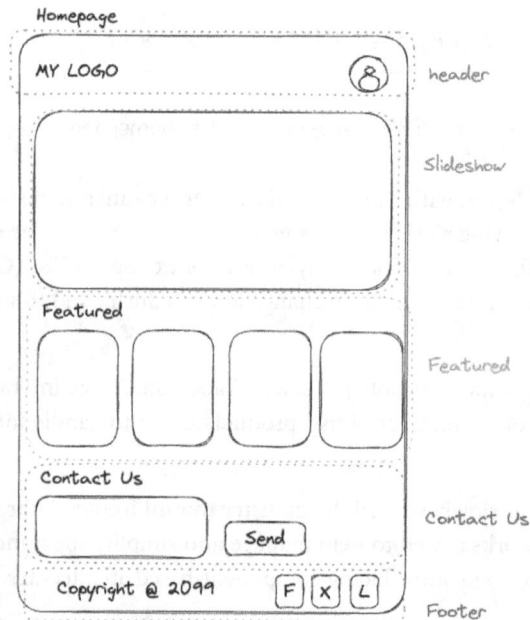

Figure 1.3: Wireframe of a dashboard divided into different sections, such as header, slideshow, featured, Contact Us, and footer

As we will see in a few minutes, the components displayed in *Figure 1.3* are just an example, as a fully defined CBD will actually break things even further all the way to the single HTML element. What this means is that not only the page is made of components, but components are made up of smaller components too.

Breaking down things into smaller units has many benefits. Let's analyze some of these characteristics:

- **Reusability**: CBD provides you with the possibility to create components that can be reused within your application. (In our example we could reuse the header, footer, slideshow, and even the featured component.)

- **Encapsulation**: Encapsulation is defined as the ability for each component to be "self-contained." All styles, HTML, and JS logic are "encapsulated" within the scope of a given component.

- **Independence**: Due to encapsulation, each component is independent and does not share (or is not supposed to share) responsibility with other components. This allows components to be used in different contexts (for example, the ability to use the feature component on a different page of the site).

- **Extensibility**: Due to the component being "self-contained" and independent, we are able to extend it with limited risk.

- **Replaceability**: The component can easily be swapped out with other components or be removed without risk.

It is clear from the preceding list that using CBD brings many benefits to the hands of a frontend developer. As we will experience in the course of this book, the ability to break an application down into small units is extremely beneficial for new developers as it allows the individual topics to be broken down and really focuses our attention on what matters the most.

Vue.js implements component-based architecture with the use of a **Single-File Component** (**SFC**). These files are denoted by the `.vue` extension and encapsulate styles, HTML, and logic (JS or Typescript) in the same file. SFC will be clearly introduced later in the book.

Atomic design

In this last section, we are going to understand how we will structure our components during the course of the book.

The folder structure of components is something that has not been standardized yet around the industry and this can differ from developer to developer.

In our case, we are going to follow what is known in the industry as "atomic design." This is described as:

The Atomic Design methodology created by Brad Frost (https://bradfrost.com/) is a design methodology for crafting robust design systems with an explicit order and hierarchy - `blog.kamathrohan.com`.

The atomic design pattern follows the same concept described in chemistry and the composition of matter. If you want to go into more detail on this subject, I suggest you read the following article: `https://blog.kamathrohan.com/atomic-design-methodology-for-building-design-systems-f912cf714f53`.

In this book, we are going to follow the hierarchy proposed in this methodology by breaking down our applications into "sub-atomic," atoms, molecules, organisms, templates and pages.

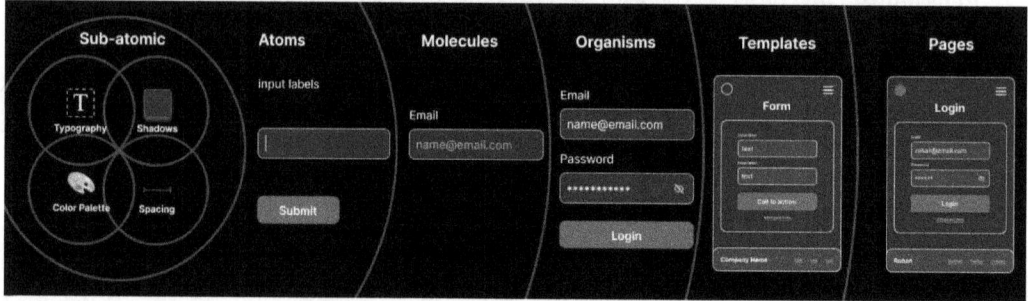

Figure 1.4: Visual explanation of the different levels offered by atomic design (source: `https://blog.kamathrohan.com/atomic-design-methodology-for-building-design-systems-f912cf714f53`)

Atomic design layers are as follows:

- **Sub-atomic**: The sub-atomic layers include all the variables and settings that will be used within the application. These are not going to be "components," but just CSS variables that will be shared globally within our application. In the sub-atomic layer, we find colors, typography, and spacing.

- **Atoms**: These are components that will define individual HTML elements, so, for example, a button, an icon, and an input text are all part of atoms.

- **Molecules**: Molecules are made up of two or more atoms or plain HTML elements. For example, an input field with a label and an error is a molecule.

- **Organism**: There are UI components that make up a standalone section that can be used on the site. For example, a login form is an organism, a slideshow is an organism, and so is a footer.

- **Templates:** These are commonly called layouts within the frontend ecosystem and are used to define a reusable structure used by multiple pages. An example could be a template with a hero image, a sidebar, main content area and a footer. This template would be used by many pages within the application and abstracting it into its own template reduces duplication.

- **Pages**: Lastly, we have pages. These are used to define our web application page or subpage. A page is going to be the place in which our data is loaded, and it will include HTML elements, organisms, molecules, and atoms.

Even if this separation may seem complicated to understand from the preceding description, we will touch base on this topic multiple times during the book and this will help you understand the main difference between the layers available.

Spend some time going through the folder structure of the application and read the components' names to try and understand how we will break up our application.

Separation of concern

So far, we have learned that the modern framework offers the ability to break up the application into small chunks called components and that there is a hierarchy within the component itself.

In this section, we are going to quickly touch base on why this hierarchy was introduced in the first place and understand how this is going to help our development.

Atomic design not only supports us in breaking up components by their visual complexity, but it also helps us to break up the application logic to create highly performance and scalable applications.

As the component definitions get more complex, so does the logic expected to be attached to it.

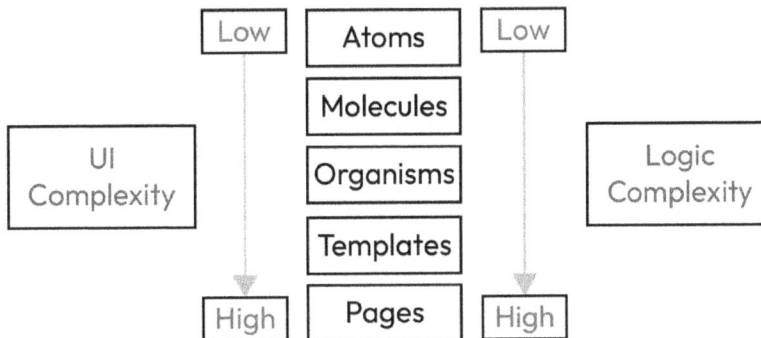

Figure 1.5: Illustration of the level of UI and Logic complexity for each layer

What do we mean by logic complexity? Logic complexity can be described as the amount of JS required for the component to function correctly.

For example, a component with low logic complexity such as a button will have very limited JS, while a more complex component such as an input field will need to handle field validation and error placement; furthermore, a page will have to take ownership of loading the data from the API, format the data to ensure that it is in the right format, and handle the visibility of its children.

In this section, we have introduced how an application is structured using the component-based architecture. We introduced the different layers that make up a component library and finally defined the advantages that this methodology when used in conjunction with a frontend framework such as Vue.js. Let's now recap the chapter in the *Summary* section.

Summary

It is now the end of the chapter and at this stage, you should have gained some knowledge of what we will achieve in this book and the methods that we will be using.

We have learned about our Companion App and what that will include. We have quickly touched upon the chapters' structure and how they will support you in your learning journey, and we have finally introduced important topics such as component-based architecture, atomic design, and the core areas of web development that are the foundation of any frontend project.

In the next chapter, we will start to learn about the foundation of Vue.js and its core fundamentals and start to give life to our Companion Application.

2

The Foundation of Vue.js

If you are reading this book, chances are that you have decided to use Vue.js as your framework of choice and there is very little reason to try to convince you not to use it. We are going to use this chapter to begin sharing details of what makes Vue.js unique and why it has become so successful.

We are first going to learn what makes Vue.js different from other frameworks; we will then move on to study Vue.js' reactivity and its lifecycles. Finally, we will learn about the component structure of Vue.js.

In this chapter, we will learn the following:

- Vue.js' reactivity fundamentals
- Understanding the Vue.js lifecycle and hooks
- Vue.js' component structure

The goal of this chapter is to provide you with information regarding Vue.js that will become the foundation of your future learning. Understanding Vue.js' reactivity will help differentiate Vue.js from other frontend frameworks and libraries, and you will learn the complete lifecycle of a Vue.js component to help you make the correct technical decisions. Finally, understanding the different ways to define a Vue.js component will prepare you for the chapters to come.

Vue.js reactivity fundamentals

Vue.js has been around for some time; the framework's first release dates back to 2014, when its creator, Evan You, a former Google developer, informed the world of its creation.

Evan's previous experience with Angular at Google gave him the knowledge necessary to build a great framework. In an interview for *Between the Wires* shortly after making the framework public, Evan said the following:

> *I figured, what if I could just extract the part that I really liked about Angular and build something really lightweight.*

Evan did not just create a lightweight framework, but he also managed to build an amazing community around it, making it one of the most loved frameworks by developers.

Vue.js has had just three major releases until now, with the latest one being a full rewrite that made Vue.js faster, smaller, and even easier to use.

Two main aspects made Vue.js so successful. The first is that its growth and adoption are driven by the community for the community. Vue.js is one of the few major frameworks not to be backed by a big company. It is fully funded by people's donations to the core team and its development is mainly driven by the community. This is shown by the focus on the development experience that is always present within the Vue.js ecosystem.

The second aspect that makes it unique is its reactivity system. The Vue.js core engine has been built to be reactive behind the scenes, making handling states with Vue.js simple and intuitive.

When we talk about reactivity in development, we refer to the ability of certain variables to automatically update when a change occurs. A simple example of reactivity outside of the developing world is offered by Excel and Google Sheets. Setting up a calculation, such as a sum of a column, will result in the total number being "reactive" to any chance that happens in the summed cells:

	A	B	C	D	E
1		Students			Students
2	Class A	5		Class A	10
3	Class B	5		Class B	5
4	Class C	5		Class C	5
5	Total	15		Total	20

Figure 2.1: Google Sheet example showing how the value of cells updates automatically

Like Excel, reactivity in web development, particularly in Vue.js, allows your variables to be dynamic and automatically change when a value it depends on changes.

Let's see a real example to understand how reactivity plays a big part in the UI framework. Let's start by seeing the behavior of vanilla JS and then see how this translates into Vue.js.

In the following code, we are going to create two variables, `firstName` and `lastName`, and then we will try to create a reactive variable called `fullName`:

```
let firstName = "Simone";
let secondName = "Cuomo";
const fullName = `${firstName} ${secondname}`;
console.log(fullName);
// output: Simone Cuomo
```

In the preceding code snippets, the full name that is printed in the console is equal to the `firstName` and `secondName` variables that we have created. What would happen if we now change the `firstName` variable to a different value? What would the value of `fullName` be?

```
let firstName = "Simone";
let secondName = "Cuomo";
const fullName = `${firstName} ${secondname}`;
console.log(fullName);
// output: Simone Cuomo
firstName = "John";
Console.log(fullName);
// output: Simone Cuomo
```

As you can see from the preceding code, the variable that is supposed to print the full name has output the incorrect value, as it has not "reacted" to the change in the `firstName` variable.

This is perfectly normal behavior; you do not want all variables to automatically react in JavaScript, as this will complicate its usage, but when it comes to the UI, having values that are updated is the expected behavior.

Say, for example, you are filling up a basket and you want the total number of items to change if you add one more item to your basket, or you would like the word count to update if you type in a limited textbox, and so on.

Let's replicate the preceding example using Vue.js:

```
let firstName = ref("Simone");
let secondName = ref("Cuomo");
const fullName = computed( () => `${firstName.value} ${secondName.
value}`);
```

```
console.log(fullName);
// output: Simone Cuomo
firstName.value = "John";
Console.log(fullName);
// output: John Cuomo
```

At this stage, your understanding of Vue.js is still limited to what experience you had before reading this book, so you are not expected to understand the preceding code yet. What we need to focus our attention on is the output that the code produces when compared to the one produced by plain JavaScript.

As the preceding code shows, the `fullName` variable changes automatically as soon as any of its dependent variables (`firstName` and `secondName`) change.

Understanding how the reactivity system works behind the scenes is out of the scope of this book, but this does not prevent us from understanding the technicalities that take place behind the scenes.

The following diagram shows what happens behind the scenes and how the reactivity actually takes place:

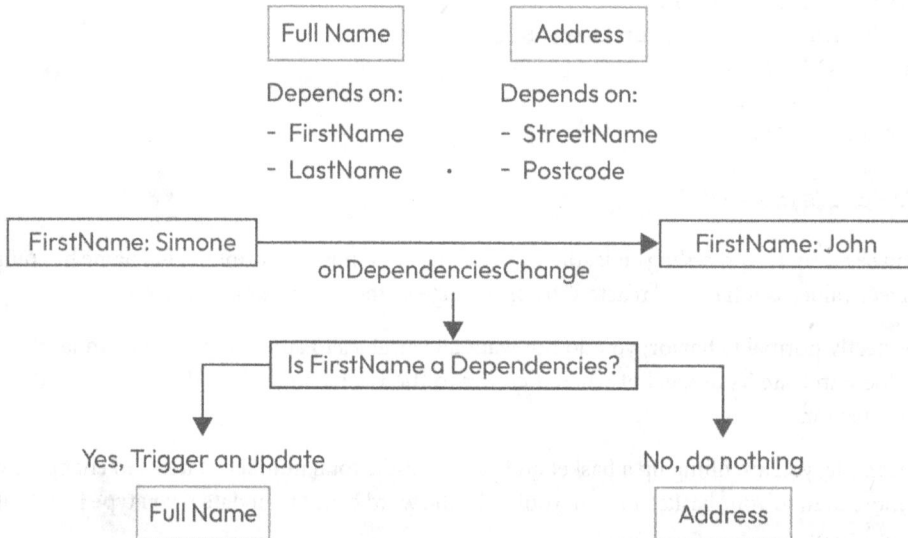

Figure 2.2: Diagram explaining the Vue.js reactivity system

This illustration is a simplified version of what happens within the Vue.js reactivity core system to make our variables dynamic. Let's break down what is happening:

1. We defined reactive variables, such as `FirstName`. Vue.js watches this variable for any change event.

2. We declared a complex variable that is dependent on other reactive variables (e.g., `fullName`).

3. Vue.js tracks a dependency tree. It creates a list of what depends on what.

4. When a change takes place in a reactive variable, the reactivity engine will trigger `onDe-pendenciesChange`.

5. Vue.js evaluates which values depend on the value that was just changed and triggers an update *only* if the value is part of their dependencies.

While reading the preceding process, you probably thought that it sounded quite familiar, and you would be correct, as the reactivity system follows the same principles offered by HTML elements such as input fields. Elements such as `<input>`, `<select>`, and many more have the ability to hold values and react when they are changed by triggering an `onChange` or similar event.

As shown in the preceding example, listening to a change event to handle data reactivity is not something unique. So, what makes the Vue.js reactivity system special? Vue.js reactivity stands out for the way it handles the dependencies tree and automatically updates variables behind the scenes. The reactivity system of Vue.js is non-obstructive and it is completely unseen by developers. Vue.js manages all the dependencies behind the scenes as part of its **lifecycles** and acts upon changes with speed and high performance.

This section introduced you to the Vue.js reactivity system, explaining how it plays a vital role in the success of the framework. We then explained, with the help of some examples, how the engine works behind the scenes. It is now time to understand how the Vue.js core engine works by taking a closer look at its lifecycles and understanding how they can be used within our application development.

Understanding the Vue.js lifecycle and hooks

As we progress into the book, our knowledge of Vue.js continues to expand. In this section, we are going to discuss the Vue.js lifecycle.

When we use Vue.js, the application goes through a defined list of steps, from creating the component HTML to gathering all the dynamic values, as well as displaying these values in the DOM. Each of these is part of what we call the **lifecycle**, and in this section, we are going to define them all and learn when and how to use them during the course of our development careers.

If you have ever tried to learn Vue.js in the past, you have probably already been exposed to the following diagram, which is available in the main Vue.js documentation:

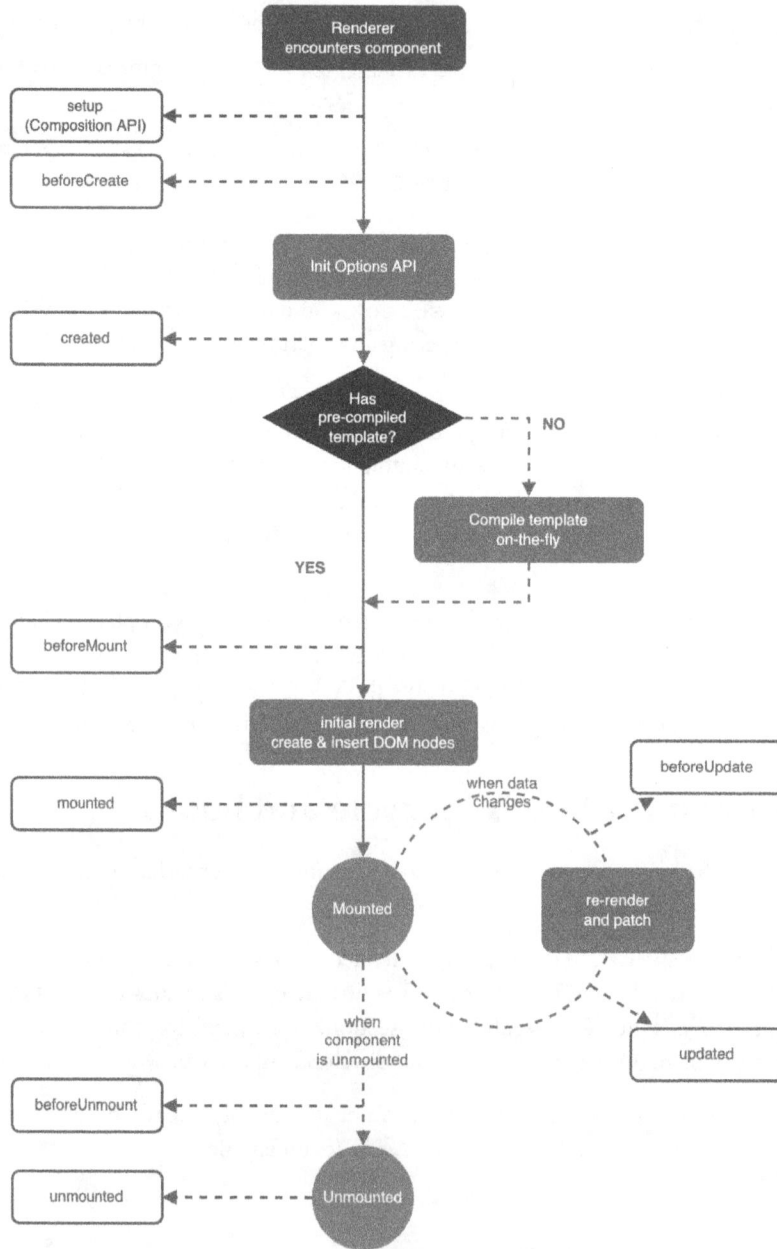

Figure 2.3: Vue.js lifecycle diagram (from www.vuejs.org)

No matter how long you have been using Vue.js, the preceding diagram will repeatedly appear in your browser history, and it will slowly be imprinted in your memory, as it is the foundation of Vue.js and a must-know in order to write clean and performant code.

As you progress in the course of this book, you will be asked to revisit different parts of the lifecycle, and you will be asked to revisit the diagram.

In the next section, we are going to review the diagram step by step and understand what it means and how this knowledge can be applied during development.

We are going to start explaining it from the top down, but we will be starting with **beforeCreate**. We are purposely leaving **setup** for later, as it is easier to understand after all the lifecycles have been introduced, even if it is the first part of the list.

The following lifecycles are progressive; this means that the end state of one of them is the beginning of the next.

beforeCreate

This lifecycle is created as soon as a component is initialized. At this stage, our component does not exist at all. The Vue.js framework has just been instructed to create it and it is triggering this hook to inform us that the component is on its way.

At this stage, nothing of the component is available, no HTML is being created, and no internal variables are set yet.

Usually, this lifecycle is used to trigger analytics logs or long async tasks.

created

At this stage, Vue.js knows about your component, and it has loaded its JavaScript logic, but it has not yet rendered or mounted any HTML.

This is the perfect stage to trigger async calls to gather some data. Triggering a slow request now will help us save some time, as this request will continue behind the scenes while our component is being rendered.

beforeMount

This lifecycle is triggered right before the HTML is appended to the DOM. There are very limited use cases for this lifecycle, as most of the pre-render actions are triggered within the **created** lifecycle.

mounted

At this stage, the component has been fully rendered, and its HTML has been attached to the DOM. If you need to complete any operation that requires you to access the DOM, this is the correct lifecycle, as the HTML is ready to be read and modified.

If you come from a non-framework background, you may think that most of the logic of your component will probably be included in this lifecycle, but you will quickly learn that due to the way Vue.js components are specified, you will rarely need access to the DOM.

beforeUpdate and update

beforeUpdate and **update** form a recursive circle that happens any time the component data or dependencies change. We already introduced this step in the previous section when we spoke about the reactivity system.

beforeUpdate is triggered as soon as Vue.js realizes that a reactive value on which the component depends has changed.

On the other hand, **update** is triggered when the value has been fully changed, and its value has been assigned to the correct DOM node and is ready in the DOM.

You will very rarely have to use these two lifecycles directly, as Vue.js provides other features, such as computed properties and watchers, to be able to handle individual changes within the component data.

beforeUnmount and unmount

At this stage, our component is no longer needed, and Vue.js is ready to remove it from the DOM. This could be due to the user navigating to a different page or any other event that would require the component to be removed from the UI.

There is very little difference in usage between **beforeUnmount** and **unmount**. This lifecycle is very useful for unsubscribing to events, such as "click" and "observers," that will result in a drop in performance if left active.

setup

As promised at the start of this section, we left **setup** for last, as it is easier to explain it after all lifecycles were covered. **setup** is not a lifecycle in itself, but it is the entry point used by the CompositionAPI (something that you will learn about a bit later in this chapter). When using **setup**, you have the ability to call and access all the lifecycles (**mounted, updated, unmounted**, and so on). You can think of **setup** as a wrapper for the Vue.js lifecycle, a single method that includes all lifecycle hooks. Composition API is going to be what we use in this book, and we are going to explain the **setup** function in much more detail at a later stage.

In this section, we have learned the basic flow of Vue.js, introducing all its lifecycles. At this stage, we should know when Vue.js component is rendered, updated, or destroyed. This knowledge will drive our development and allow us to make the correct choices to make our application performant. In the next section, we are going to see how to introduce Vue.js component syntax, and we will also learn how to make use of the preceding lifecycles.

Vue.js component structure

Components are the basis of the Vue.js framework. They are the building blocks required to create an application using this framework. As was previously explained, a component can be as small as a simple button or as large as a full page.

No matter their size, all components are built using the same syntax and structure. In this section, we are going to learn the different forms of syntax available to write components and learn about the different sections that make up a Vue.js **single-file component (SFC)**.

Single-file components

SFCs are specific to Vue.js and can be found in Vue.js projects with the extension `.vue`. These files are composed of three main sections: **template**, **script**, and **style**:

```
<template></template>
<script></script>
<style></style>
```

The Vue.js compiler takes the preceding three sections and splits them up into individual chunks during build time. We are now going to explain each of them in this section. We will cover the SFC section in the following order:

1. Template

2. Style

3. Script

The <template> tag

The first section is `<template>`. This section includes the HTML hosted by our component. So, if we take an example of an extremely simple button, the template will look like this:

```
<template>
    <button>My button</button>
</template>
```

In contrast to React, the HTML of a Vue.js component is plain **HTML** and not **JSX**. As we will learn in the course of the book, Vue.js provides some handy tools to simplify the content of our HTML.

> **Important note**
>
> As for template writing styles, it is possible to write your HTML with different methods, such as render functions, or by writing it in JSX (with the correct loader), but these two methods are for specific uses and are not expected within the Vue.js ecosystem.

The <style> tag

The next section available is `<style>`. This section will include the styles associated with our component using plain **CSS**. If you are not completely new to Vue.js, you may have realized that what I just said is not fully true, as using a `<style>` tag in a component does not *scope* the style to that specific component.

Before we move on, let's explain what it actually means for styles to be scoped and how to achieve this in our Vue.js application.

When we use a simple `<style>` tag, as shown in the preceding example, our style will leak to the rest of the application. Anything we declare in the tag will be global unless we scope it with CSS:

```
<style>
p {
  color: red;
}
</style>
```

Writing the preceding style in Vue.js is permitted and is even suggested for performance and maintainability reasons. The problem is that the preceding declaration will change the color of our paragraph to red in the whole application and not just in the component in which it has been written.

Luckily for us, Vue.js has a handy tool to use in the case where we would like our component to be fully scoped, making sure that no style bleeds and breaks the rest of the app. To do so, we need to add an attribute called `scoped` to our `<style>` tag:

```
<style scoped>
p {
  color: red;
}
</style>
```

With this new attribute, our styles will be locked to the component in which they are defined, and they will not affect the rest of the application. We are going to learn more about when it is best to use these two methods when building our Companion App.

The <script> tag

The next section available within an SFC is the `<script>` tag. This tag will include the component JavaScript logic, from the properties that are accepted for the component to the private data used to define the component logic, all the way to the actual methods needed for the component to function properly.

Just a few years ago, when Vue.js' major version was still 2, components were mostly defined using a syntax called **Options API**. Even though other methodologies were available, this was the main way to write a Vue.js component.

With the release of Vue.js 3, a new method of writing components was created. This is offered alongside the existing Options API, and it offers better TypeScript support, improved techniques to reuse logic, and flexible code organization. This method is called **Composition API**.

> **Important note**
> Composition API is also referred to as **Script Setup**.

At this moment in time, neither of the two methods is officially preferred over the other; this is also emphasized by the Vue.js official documentation, which currently showcases all its tutorials and examples using both methodologies, offering the option to switch between the methods:

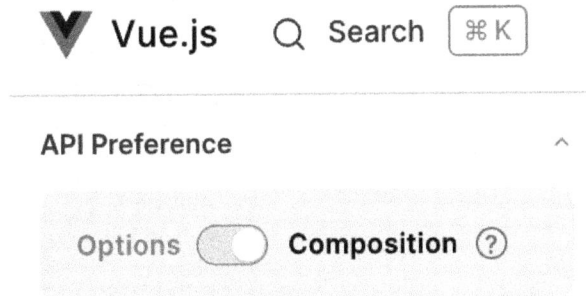

Figure 2.4: Vue.js official documentation for API preference switch

The content of this book and its Companion App are going to be written using Composition API. This decision was made for two main reasons:

- Due to Vue.js 2's history, the web is full of resources that focus on Options API but less so on the new Composition API syntax

- Evan You (the creator of Vue.js) has predicted (more than once) that, in the long run, Composition API will take over and become the standard

Because I am a strong believer that extra knowledge does no harm, in this section, we are going to learn how to define the component in both syntaxes, including Options API. Knowing both methods can help you build a strong foundation to support your learning of the Vue.js framework.

Options API versus Composition API – Two sides of the same coin

Under the hood, both methods are actually going to produce a very similar output, with Composition API producing slightly more performant code. Nevertheless, the syntax differences and benefits that these methods bring are quite different and can make a big change depending on your habits.

The first and main difference between Composition API and Options API is in the way the `<script>` section of your code is broken down. As I mentioned before, both syntaxes will offer the same features, so this means that we can declare props data, compute, and methods, as well as access all the lifecycles in both methods, but the way we do so differs.

The differences are as follows:

- **Composition API**: Code broken down by functionality

- **Options API**: Code broken down by Vue.js options

Let's look at an example to clearly define the difference between both methods.

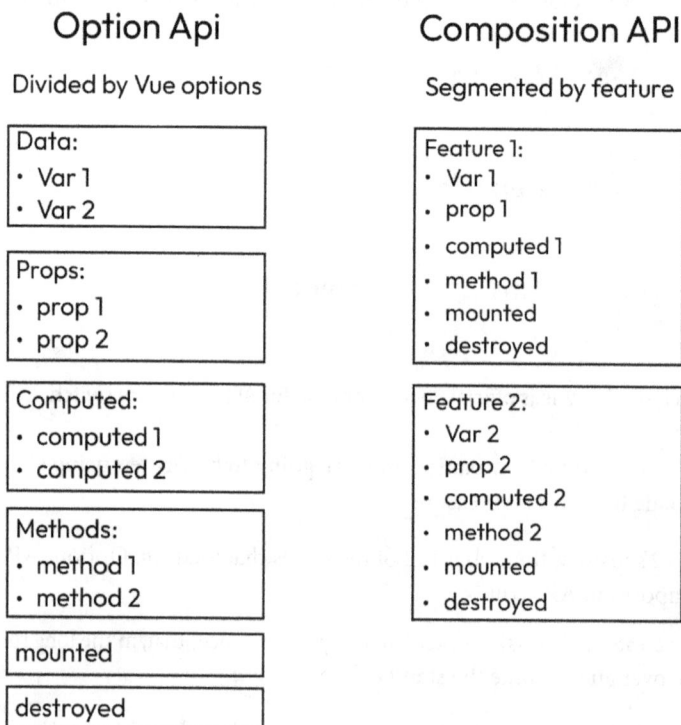

Option Api

Divided by Vue options

```
Data:
• Var 1
• Var 2
```

```
Props:
• prop 1
• prop 2
```

```
Computed:
• computed 1
• computed 2
```

```
Methods:
• method 1
• method 2
```

```
mounted
```

```
destroyed
```

Composition API

Segmented by feature

```
Feature 1:
• Var 1
• prop 1
• computed 1
• method 1
• mounted
• destroyed
```

```
Feature 2:
• Var 2
• prop 2
• computed 2
• method 2
• mounted
• destroyed
```

Figure 2.5: Comparison between Composition API and Options API for breaking down code

As shown in the preceding diagram, in Options API, the code does not take into consideration the actual component requirements and logic, but it is sliced vertically using Vue.js options: **Props**, **Data**, **Methods**, **Computed**, **mounted**, and so on.

On the other hand, Composition API takes a different approach by breaking down the component by its technical output. This allows us to create a section for Feature 1, a section for Feature 2, and so on.

The second difference is associated with TypeScript support. This is the main reason that led the Vue.js core team to decide to create The Composition API during the Vue.js 3 rewrite. Options API offers very basic TypeScript support, and this has prevented many developers from joining the Vue.js ecosystem.

We have reached the end of this section, and it is time to clearly say which method is better, but, unfortunately, the answer is that it depends.

Because both syntactic sugars compile in the same code, the decision really goes back to coding preferences. Options API provides more structure, and it can, therefore, be more helpful at the start of your career when your experience in creating a component is still limited, while Composition API, with the added TypeScript support and greater flexibility in code partitioning, can be a very strong tool to improve code readability in big applications.

Sample components

At this stage, we have learned enough about the foundations of Vue.js to be ready to introduce some sample components and see the framework in action.

We are going to look at an example of an Atom. In our case, it is a simple icon component. This sample component is going to exhibit the following features:

- It is going to accept a couple of properties (`size` and `name`)
- It is going to include some style
- It is going to dynamically load the icon

The component will be called using the following HTML:

```
<vfb-icon name="clog" size="small" @click="doSomething />
```

As previously mentioned, in this section, I am going to show the components utilizing both writing methods; however, later in the book, we will just write components using Script Setup (Composition API).

> **Important note**
> Please note that we will cover all of this again in more detail later in the book. This is just a quick introduction to Vue.js components.

An Atom component using Options API

Let's first see how this component looks as a Vue.js component using Options API:

```
<template>
  <img
    :src="iconPath"
    :class="sizeClass"
  />
</template>
<script>
  export default {
    name: 'vfb-icon',
    props: {
```

```
      size: String,
      name: String
    },
    computed: {
      iconPath() {
        return `/assets/${this.name}.svg`;
      },
      sizeClass() {
        return `${this.size}-icon`
      }
    }
};
</script>

<style scoped>
.small-icon {
    width: 16px;
    height: 16px;
}
.medium-icon {
  width: 32px;
  height: 32px;
}
.large-icon {
  width: 48px;
  height: 48px;
}
</style>
```

Now let's break down all the sections, starting with `<template>`, which hosts the HTML for our component; in this case, this is a native `<img>` element. This component has a few attributes being passed to it. The first two are native attributes: `src` and `class`:

```
:src="iconPath"
:class="sizeClass"
```

These attributes are declared a little bit differently than you are used to in HTML, as they are preceded by `:`. When an attribute has this syntax, it means that its value is dynamic and that the values (in our case, `iconPath` and `sizeClass`) are going to be evaluated as a JavaScript variable and not actual strings.

> **Important note**
> Please note that you can write plain HTML in Vue.js, and using the dynamic variables is not a requirement for the framework but just a feature to make the attributes dynamic.

Next up, let's move to the logical part of the application, the `<script>` section. Here, we start by declaring the name of our component:

```
name: 'vfb-icon',
```

> **Important note**
>
> It is good practice for all Vue.js components to be formed of two words. This will ensure that the component does not clash with native HTML elements.

Next, it is time to declare the properties. Properties are values that are accepted by our component when it is initialized. This is an existing concept in development, as all HTML elements accept attributes such as `class`, `id`, and `value`. These properties make our components reusable and flexible. In our example, we declared two different properties: `name` and `size`. These will be passed down when the component is called, just as if they were native HTML attributes:

```
props: {
  size: String,
  name: String
},
```

This example shows a basic configuration for a property, in which we just define its type, but, as we will later see in the course of the book, `props` can have different configurations, such as validation, default values, and requirements rules, to state whether they are required or not.

The next piece of code is where we declare our dynamic properties. For our component to function correctly, we need a path and a class defined as `iconPath` and `sizeClass`. These values are going to be dynamic due to the fact that they include the dynamic properties and will be declared using something called computed properties:

```
computed: {
  iconPath() { return `/assets/${this.name}.svg`; },
  sizeClass() { return `${this.size}-icon` }
},
```

The computed properties allow us to declare values that are reactive (remember the reactivity chapter earlier on in the book) and can make use of our entire component logic; in this case, we just used the props, but we could have used a collection of values and external logic to create a new value.

> **Important note**
>
> Please note that when using Options API, you have to use the `this` keyword to be able to access variables within the component, such as `props` and `computed`. In our example, we used it to access properties by using `this.name` and `this.size;`.

Our last section is `<style>`:

```
<style scoped>
  .small-icon {
    width: 16px;
    height: 16px;
  }
  .medium-icon {
    width: 32px;
    height: 32px;
  }
  .large-icon {
    width: 48px;
    height: 48px;
  }
</style>
```

This is quite simple because this example does not include anything different than you would normally see in plain CSS. As mentioned in a previous chapter, we can add the attribute scoped to our style to ensure that their style does not bleed from our component.

In the preceding example, you can see (in practice) how Options API divides our component into sections. In our case, we have `props`, `name`, `computed`, and `methods`:

```
export default {
  name: '',
  props: {},
  computed: { },
  methods: { }
}
```

An Atom component using Script Setup

It is now time to look at the same component but written using the `<script setup>` syntax:

```
<template>...</template>
<script setup>
  import { computed } from 'vue';
  const props = defineProps({
    size: String,
    name: String
  });
  const iconPath = computed( () => {
    return `/assets/${props.name}.svg`;
```

```
  });
  const sizeClass = computed( () => {
    return `${props.size}-icon`;
  });
</script>

<style scoped>...</style>
```

As you can clearly see, the preceding example omits the `<template>` and `<style>` tags. These have been omitted because they are identical to the Options API counterpart. In fact, as we have already mentioned, the difference between these two methods only affects the logical part of the component, which is `<script>`.

The first line of our component is `import`:

```
import { computed } from 'vue';
```

In contrast to Options API, where all the options were already available to us when using `<script setup>`, we have to import each individual Vue.js method from `'vue'`, just as we did in the previous code for `computed`.

Next up, we are going to see how properties are defined in Composition API:

```
const props = defineProps({
  size: String,
  name: String
})
```

Properties are one of the few options to have a verbose declaration while using `<script setup>`. In fact, to be able to declare them, we need to make use of a compiler macro called `defineProps`. Macros do not need to be imported, as they are just going to be used by the compiler and removed from the code. If you have ever used TypeScript, you will be familiar with this approach.

Next up, we have `computed`:

```
const iconPath = computed( () => {
  return `/assets/${props.name}.svg`;
});
const sizeClass = computed( () => {
  return `${props.size}-icon`;
});
```

Declaring the properties of `computed` is very similar to Options API but with two small differences:

- The logic of the properties of `computed` need to be passed as a callback to the computed method imported from `'vue'`

- The `this` keyword is not available anymore, and we can access variables directly

This is as far as we will go in terms of explaining the differences between Options API and Composition API for now. We will cover Script Setup (Composition API) in more detail later in the book. If you are extremely new to Vue.js, this section probably included lots of new syntax, and it was a little hard to grasp, but as soon as we start to build our Companion App, as you gain knowledge with Vue.js and its syntax, things will quickly make more sense to you.

In this section, we have started to learn how Vue.js components are defined and the different sections that form an SFC. We then concluded the section by covering a sample component in both syntactic sugars in detail.

Summary

We have now reached the end of this fairly theory-heavy chapter, and this was required for us to get started with our app-building process. In this chapter, we have learned what makes Vue.js different from the other frameworks by analyzing its reactivity system. We then broke down the composition of a Vue.js SFC, also known as a `.vue` file, and we walked through a Vue.js component's lifecycle by analyzing all the different lifecycle hooks available within the framework.

In the middle of the chapter, we learned the main differences between Composition API and Options API by exploring them using sample components.

In the next chapter, we will start to learn Vue.js by starting to build our Companion App. This will be the beginning of your long journey from being a complete beginner to an experienced Vue.js developer.

Part 2:
Understanding the
Core Features of Vue.js

Part 2 of this book focuses on Vue.js and its core features. We will build the core of the Companion App one step at a time, while continuing to expand our knowledge of the Vue.js framework.

This part contains the following chapters:

- *Chapter 3, Making Our HTML Dynamic*
- *Chapter 4, Utilizing Vue's Built-in Directives for Effortless Development*
- *Chapter 5, Leveraging Computed Properties and Methods in Vue.js*
- *Chapter 6, Event and Data Handling in Vue.js*
- *Chapter 7, Handling API Data and Managing Async Components with Vue.js*

3
Making Our HTML Dynamic

The theoretical chapters have now come to an end. It is time to start building our Companion App and learn about Vue.js. As we build our application, one step at a time, we will also learn about Vue.js. The idea of this approach of learning by doing is most effective when you follow along and build the application alongside me.

To help you grasp complicated topics and ensure that you have learned the basics of Vue.js, you will also be asked to complete some extra tasks that can either be applied to the Companion App or used as a standalone project.

In this chapter, we'll cover the following topics:

- Building your first Vue project
- Creating our first components
- Introducing properties
- Learning about Vue.js reactive data with Refs and Reactive

The goal of this chapter is to introduce you to the basics of Vue.js. You will learn how to create a project from scratch to understand how a Vue component is structured. At the end of the chapter, you will be able to create Vue components, define attributes using props, and handle private state with Refs and Reactive.

Technical requirements

From this point onward, all chapters will require you to check out a specific version of the code from our repository. We installed our repository back in *Chapter 1*, when we downloaded the application and ran it for the first time.

Using that same repository, which is available at `https://github.com/PacktPublishing/Vue.js-3-for-Beginners`, we can jump between chapters using the various branches. Having one individual branch per chapter ensures that our starting point matches, preventing possible code issues or missing parts that would make the learning complicated.

In this chapter, the branch is called CH03. To pull this branch, run the following command or use the GUI of choice to support you in this operation:

```
git switch CH03.
```

Do not forget that to run the application after switching the branch, you have to ensure all dependencies are installed and run the dev server. This can be achieved with the following two commands:

```
npm install
npm run dev
```

Please note that we won't need the repository until the *Creating our first component* section.

Building your first Vue.js project

The time has come to start and build our companion application. If you are already familiar with Vue.js and how you create a new project using it, you can skip this section, pull the code from our repository, and start building the application in the next chapter.

When we create a new project using the `vue create` command, which we will see soon, we use the Vite build tool behind the scenes. Until recently, the best build tool was Webpack, and all frameworks, including Vue 2, used it to build their applications. But things have now changed, and Vite has taken over due to its no-config approach and extremely fast development server.

On its official site, Vite is described as follows:

> *"Vite (French word for "quick", pronounced /vit/, like "veet") is a build tool that aims to provide a faster and leaner development experience for modern web"*

Vite was created by Evan You (yes, the author of Vue) in an attempt to improve the development experience. Vite has been around for just a few years, but it has quickly gained popularity due to its low configuration and fast development server.

Vite, like Vue.js, is fully open source, and it also supports all major frameworks out of the box. Creating a project with Vite is quite simple, all you need is an IDE and a terminal.

Vue CLI GUI

You have probably heard that Vue CLI offered a visual tool that helped you manage the Vue application. Unfortunately, that project was linked to Webpack and has yet to be imported to Vue 3 and Vite.

Create Vue command

Because it is not possible to create a project in an existing folder, we are not able to re-use our previously downloaded application. We are going to complete this step in a different folder to ensure that you are able to get a project started from scratch. In the next chapter, we will pull the code directly from the companion application repository.

To create a new project with Vue, we first need to access the folder in which the project will be created. Please be assured that the CLI will create a new folder for your project, so you do not need to create a folder manually now, but just access the main folder in which the project should live. For example, I like to create all my projects in the Document folder, so I am going to access it like this:

```
// Mac users
cd ~/Documents/
// Windows users
cd %USERPROFILE%/documents
```

Now that we are in the correct folder, we can call the Terminal command required to create a new Vite project:

```
npm create vue@latest
```

Running the preceding command will generate a request to install the create-vue package:

```
zelig880@Simones-MacBook-Pro Vue.js-for-Beginners % npm create vue@lates
Need to install the following packages:
  create-vue@3.7.4
Ok to proceed? (y) y
```

Figure 3.1: The installation message triggered by the create vue command

To successfully install the project, we need to press *y* and proceed with the installation.

After a few seconds, the CLI will start and ask us questions that will help it scaffold the project to align with our needs:

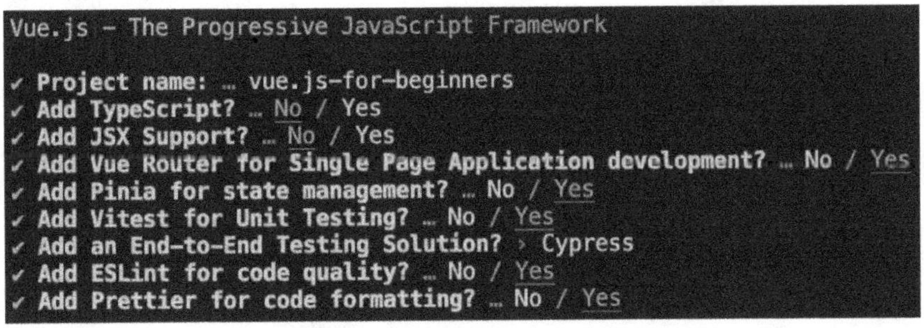

```
Vue.js - The Progressive JavaScript Framework

✔ Project name: … vue.js-for-beginners
✔ Add TypeScript? … No / Yes
✔ Add JSX Support? … No / Yes
✔ Add Vue Router for Single Page Application development? … No / Yes
✔ Add Pinia for state management? … No / Yes
✔ Add Vitest for Unit Testing? … No / Yes
✔ Add an End-to-End Testing Solution? › Cypress
✔ Add ESLint for code quality? … No / Yes
✔ Add Prettier for code formatting? … No / Yes
```

Figure 3.2: The Vue CLI questions

As you can see, Vue projects come with a nice set of presets that help it to create a strong foundation for your next Vue project. Vue CLI provides options for the following settings:

- **Project Name**
- **Typescript**
- **JSX support**
- **Router**
- **State Management**
- **Unit tests**
- **End to End tests**
- **Code quality**
- **Code formatting**

The choices for these settings are completely up to you, and you should follow your personal needs and requirements. The ones that you see in *Figure 3.2* are the settings that I have used to create the Companion App that we will use in the next chapter.

After pressing *Enter* and waiting for a few seconds, we should get some information about how to run our project. This requires us to access the folder, install the required packages, and run the development server.

First, let's navigate to the folder that was created as part of our Vue project initialization, that is equal to your project name:

```
cd "vue-for-beginners"
```

Then, install all the packages required for the project to function. I have used npm in this instance, but you can use Yarn or PNPM:

```
npm install
```

Last, we just need to run this command to run the development server:

```
npm run dev
```

After these two commands, in less than a second you should see the following message displayed in your console:

Figure 3.3: Vite output of a successful run of the Vue development environment

Vite projects act differently from the previous Webpack projects that were run on port 3000 and run on port 5173. The local URL will be displayed in the console, as shown in *Figure 3.3*.

In our case, accessing the browser on `localhost:5173` will show the following website:

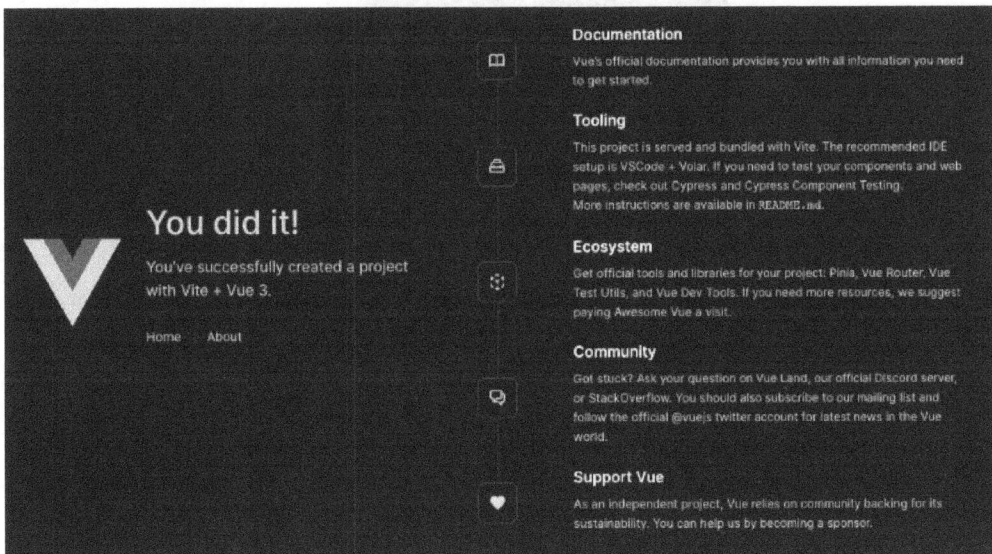

Figure 3.4: Welcome page of a newly built Vue project

Congratulations on creating your first Vue project. This is going to be the first of many projects.

Vue project folder structure

In this section, we are going to quickly cover the structure of a new Vue project.

When a new project is created, it comes with a well-defined structure that can be used as a strong foundation for future development:

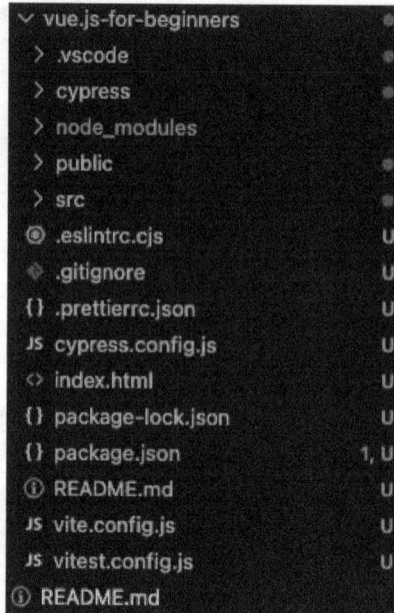

Figure 3.5: Folder structure of a newly created Vue project

We are going to explain the different folders and files to help you find anything you need from your new Vue project. We are going to do this in no particular order.

Root folder

The root folder of a Vue project includes a few configuration files. These are preset and pre-generated by the Vue creation package and do not need any further attention for the application to run smoothly. During the course of this book and your career, you will slowly be exposed to each of these configuration files and learn about their various options.

No touch zones

There are a couple of folders, such as `.vscode`, `node_modules`, and `dist`, that are what I call "no touch" folders. You may already be aware of these folders as they are created and managed by tools and software that you may have already used, such as Visual Studio Code, npm, or Vite, and should not be modified manually.

Public

The content of the public folder is going to be copied directly into the output folder after the project is built. This folder is very rarely touched by a developer but is very useful when there are files that are needed in the build output and not part of the Vue compilation. Example files for this folder are favicon and service worker.

Cypress

As shown in the installation guide, the newly created project comes with a preset **end-to-end** (**E2E**) testing framework using the tool of your choice. In our case, I selected Cypress and the CLI has created a folder and a sample test for me ready to be used.

SRC

This is where your source code lives. This is the main content of our application, and it is where you will spend most of your day-to-day work. Due to the importance of the folder, we are going to see its content and make sure we know how its files are structured:

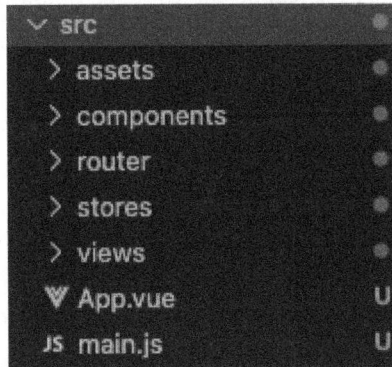

Figure 3.6: Content of the SRC folder from a newly created Vue project

As before, let's start from the root of the folder. This includes two files, `main.js` and `App.vue`. `Main.js` is the entry file of our application. This file is used to add new packages to our Vue instance and to load and set up global plugins, composable (function that leverages Vue's Composition API to encapsulate and reuse stateful logic that we will introduce later in the book) and components. Next, we have App.vue. This is the first Vue entry point and is the component that will load and handle the rest of the Vue application.

Next up, we have the `assets` folder. This folder is used to load any assets, such as images, PDFs, and videos. The content of this folder is also copied in the output artifact of our build.

Further down the list, we have the `components` folder. This folder contains not only the set of components already available within the app, but also the `__tests__` folder, which includes our unit tests.

The next two folders are `router` and `stores`. As the name suggests, they include the `vue-router` and the `Pinia` store code respectively. These are two core packages provided by the Vue.js core team and will be covered in detail in *Chapter 10* and *Chapter 11*. Vue-router will be used to create navigation routes for our clients and help us manage our growing application, while Pinia will be used to create and manage data within the application.

Last but not least we have `views`. If you have had time to investigate this folder, you would have noticed that it contains simple Vue components. The reason for this folder is to separate simple component units (the ones stored in the `components` folder) actual routing pages. Making this separation helps to keep the code clean and delineate the routing of the application.

> **Your private playground**
>
> Even if the app we just created is not needed for the rest of the book, it could be useful to keep and use it as a playground to practice the topic that we will cover in the course of the book.

We have concluded our Vue project explanation, and you should now have the knowledge required to create a new Vue project from scratch. You should also know a little bit about Vite and be able to navigate the folder structure of a newly created Vue project. In the next section, we will dive into the code and start to build our first Vue.js component.

Creating our first component

As you may remember from the first chapter, in which we introduced the Companion App, we are planning to build a clone of the social application called X (former Twitter). To start our building journey in style, we are going to build the most iconic component of the application, a post.

Figure 3.7: Example of a X.com post component

In this chapter, we are going to learn how to switch between the different branches within the book repository. We are then going to create our first SFC (**Single File Component**) component using the template knowledge we gained in the previous chapter, and we will finish by adding some dynamic content to the component by introducing two Vue reactivity functions, `Refs` and `Reactive`.

Creating Post.vue

Back in *Chapter 1*, we mentioned that we were going to break down components into different layers (atoms, molecules, organisms, and so on) and `SocialPost.vue` is going to be part of the molecules layer.

So, let's create a folder called `molecules` in the `component` folder and then add a file called `SocialPost.vue`. Once you've done this, your folders should look like this:

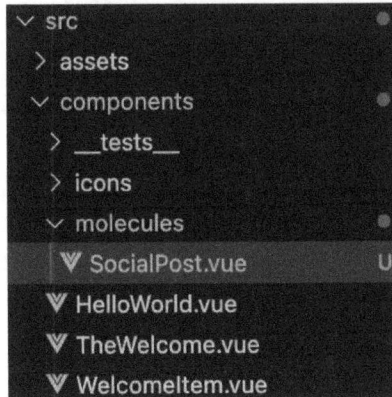

Figure 3.8: File tree of the src folder of the companion app

There are two things to notice about the new file we have created:

- The name is made up of two words. This is not just to provide more context, but also because one letter component, like `post.vue`, are discouraged due to possible future collision with native HTML components like `<button>` or `<table>` (eg. If a new HTML element called `<post>` is introduced in future HTML releases it could clash with our custom component)

- Component names are written in PascalCase, a naming convention in which each word that makes up the variable is capitalized.

As our file is empty, let's open it and start to create the basic structure of a Vue component by adding the `<template>`, `<script>`, and `<style>` tags:

```
<template></template>
<script setup ></script>
<style lang="scss"></style>
```

This is going to be our standard Vue starter template. It defines an empty template where we will encapsulate our **HTML**, a script tag with the `setup` attribute, which allows us to write JavaScript logic using the Composition API, and a `style` section, in which we select **SCSS** as our preprocessor.

We are now going to define the HTML and the CSS required for our post to be displayed. This is going to be a very simple design for now; we will add more in the course of the book.

Our first draft of the component is going to include a header image, the name of the user followed by the ID of the user, and the post's description. All of this is also going to include some basic styles. Let's look at the code:

```
<template>
<div class="SocialPost">
  <div class="header">
```

```
    <img class="avatar" src="https://i.pravatar.cc/40" />
    <div class="name">Name of User</div>
    <div class="userId">@userId</div>
  </div>
  <div class="post">This is a dummy post</div>
</div>
</template>

<script setup >
</script>

<style lang="scss">
.SocialPost{
  .header {
    display: flex;
    align-items: center;
    margin-bottom: 8px;;
  }
  .avatar {
    border-radius: 50%;
    margin-right: 12px;
  }
  .name {
    font-weight: bold;
    margin-right: 8px;
    color: white;
  }
}
</style>
```

As you have probably noticed, the preceding component does not have anything special. There's no `script` tag, no special tag in the HTML, and nothing special in the CSS, but it is still a perfectly normal Vue component.

Let's highlight a few important aspects of this component:

- `<div class="SocialPost">`: It is good practice to always assign a class equal to the component name, that in our case is SocialPost, to the root element of the component. This will help us scope the style without needing to use the `scoped` attribute.

- `<style lang="scss">`: In our examples, we are going to use **SCSS**. This is specified here. As you will see in the following section, this needs to be configured in our Vite project. You do not need to add a pre-processor, but I am adding one to show you how to add and use it in case you are used to writing your style with one.

- .SocialPost{: We can use the class we attached to the component name to scope our CSS by wrapping all the styles into it. This will ensure that our style will not bleed into other components.

Now that the component is ready, it is time to test it out. To do so, we need to load the component somewhere in the application. We can do so by loading the component in TheWelcome.vue.

To successfully load a Vue custom component, we need to complete two simple actions. First, we need to import the component, and second, we need to call it in the HTML as if it was a native component.

To load the component, we import it like a normal JavaScript file:

```
<script setup>
  import SocialPost from './molecules/SocialPost.vue'
</script>
```

Now that the component is loaded, we can simply use it in our HTML like so:

```
<template>
  <SocialPost></SocialPost>
</template>
```

Now that our component has been fully developed and loaded, it is time to try it out. To do so, let's run our application using the npm run dev command. Now, access the localhost site shown in the terminal (http://localhost:5173/).

Unfortunately, the browser output is not what we expected; we are presented with an error:

```
[plugin:vite:css] Preprocessor dependency "sass" not found. Did you install it? Try
`npm install -D sass`.

/Users/zelig880/Documents/Vue.js-for-Beginners/src/components/molecules/SocialPost.
vue

    at loadPreprocessor (file:///Users/zelig880/Documents/Vue.js-for-Beginners/node_modules/vite/dist/
    at scss (file:///Users/zelig880/Documents/Vue.js-for-Beginners/node_modules/vite/dist/node/chunks/
    at compileCSSPreprocessors (file:///Users/zelig880/Documents/Vue.js-for-Beginners/node_modules/vit
    at compileCSS (file:///Users/zelig880/Documents/Vue.js-for-Beginners/node_modules/vite/dist/node/c
    at process.processTicksAndRejections (node:internal/process/task_queues:95:5)
    at async TransformContext.transform (file:///Users/zelig880/Documents/Vue.js-for-Beginners/node_mo
    at async Object.transform (file:///Users/zelig880/Documents/Vue.js-for-Beginners/node_modules/vite
    at async loadAndTransform (file:///Users/zelig880/Documents/Vue.js-for-Beginners/node_modules/vite
    at async viteTransformMiddleware (file:///Users/zelig880/Documents/Vue.js-for-Beginners/node_modul

Click outside, press  Esc  key, or fix the code to dismiss.
You can also disable this overlay by setting server.hmr.overlay to false in vite.config.js.
```

Figure 3.9: Error message displayed by Vite

Luckily for us, the error is expected. As I mentioned before, SASS requires us to do further configuration. I wanted to show you how Vite would react if something was misconfigured. As displayed in the error message, Vite noticed that we are using SASS and is also providing us with the command required to install it. So, let's go and run this command in the terminal:

```
npm install -D sass
```

After running this command and refreshing the browser, our application should now display our component:

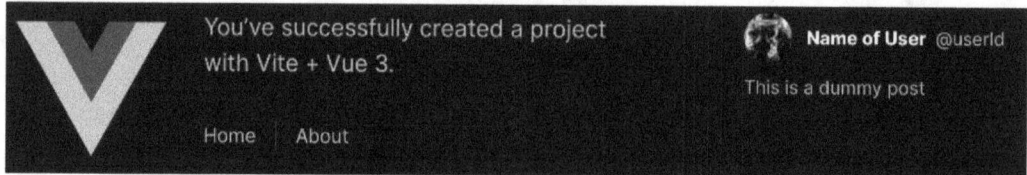

Figure 3.10: Vite welcome screen displaying a newly created custom component

Congratulations! You have just written your first working Vue component. This was just a small step, but it is important to celebrate every achievement.

As you may have noted from the file that we just created, Vue.js allows you to write simple components made up of just HTML and CSS. This is a great way to slowly get started with Vue using your existing development knowledge.

> **Your turn**
>
> Try to add another component of your own, for example, to ensure that you have understood how a component is created and added. You should try and create a static footer for our application.

In this section, we learned how to create and load a Vue component, we debugged our first issue with Vite and learned how to install new plugins, and finally, we found a new way to scope style to our component by wrapping its CSS. In the next section, we are going to learn how to make our component dynamic by introducing a feature called **properties**.

Introducing properties

As you have noticed, the component we created in the previous section is static and could not be used in a real-life application because it would always display the same information and not the actual post.

In the following section, we are going to add some dynamic features to our component. To ensure each topic is understood fully, we will add a small feature in each section and ensure that we take enough time to reiterate the features over the course of the book.

In this section, we are going to change the structure of our post component by exposing properties. Properties are simply attributes that are exposed by the component to allow users to customize its behavior or style.

If you have ever used HTML, you are probably already familiar with Vue.js props. Many native HTML elements have attributes that are used to modify components, such as an `<input>` tag as an attribute of `Type` to change its look, a `<textarea>` tag as an attribute of `column` and `rows`, and an `<img>` tag as an attribute of `src` to define the URL for its image.

Vue.js properties (usually referred to as props) allow you to define this attribute in our component, enabling us to turn our static component into dynamic and flexible building blocks.

In this section, we are going to pick our previously created post component and expose some Vue.js properties to allow us to use it multiple times with different values.

After revisiting our post component, it seems clear that the following variables should be changed to be dynamic entries:

- `Username`: Twitter username
- `UserId`: Twitter ID
- `AvatarSrc`: Source of the avatar image
- `Post`: The content of the post

Declaring props in Vue.js

The first step required for us to use props is to declare them in the component. Declaring a property means defining its name and its type. To do so, we can use a compiler macro used `defineProps`:

```
<script setup >
  const props = defineProps({
    username: String,
    userId: String,
    avatarSrc: String,
    post: String
  });
</script>
```

As shown here, the `defineProps` macros accept an object with our properties. In our case, these are all `String`, but other types, such as `Number`, `Object`, `Array`, and `Boolean`, are also accepted.

When we declare a property, we inform the component and its users that this component is happy to accept this extra data.

It is now time to learn how to access these properties in our SFC.

Accessing properties in a Vue.js SFC

Properties can be accessed in multiple places. They can be read directly in the HTML as a play string, they can be used within an HTML element declaration, or they could be used in the `script` tag as part of our component logic.

All these methods have a different syntax, but even if it seems like a lot, it is going to be quite easy to remember because it is consistent with the Vue writing style.

First, we are going to learn how to use a prop as plain text. This is done using two curly braces: `{{ props name }}`. Applying this to our template will produce the following code:

```
<template>
<div class="SocialPost">
  <div class="header">
    <img class="avatar" src="https://i.pravatar.cc/40" />
    <div class="name">{{ username }}</div>
    <div class="userId">{{ userId }}</div>
  </div>
  <div class="post">{{ post }}</div>
</div>
</template>
```

As you can see, the values of `username`, `userId`, and `post` are not hardcoded anymore and it is now using properties under the hood.

Next up, we are going to learn how to use variables in our template. I used the generic word variable instead of props because this notion applies to all variables and not just props. To use a dynamic value in our template, for example, as an HTML element attribute, we just need to prepend the attribute with the symbol `:`. So, in our scenario, the image attribute `src="..."` will become `:src="avatarSrc"`.

Our `<img>` element looks like this:

```
<img class="avatar" :src="avatarSrc" />
```

Prepending an attribute with `:` informs Vue that the value is not a plain string but an actual JavaScript variable. So, in the proceeding example, the class is evaluated as a string, but the value of `src` is not going to be literally `avatarSrc` but the JavaScript variable associated with that name.

Lastly, we are going to learn how to access properties within the `script` tag. This is achieved by using the return value of the `defineProps`.

Let's put into practice what we just learned by trying to use our properties by logging the value of `username` when the component is mounted. The code should look like this:

```
<script setup >
import { onMounted } from 'vue';
```

```
const props = defineProps({
  username: String,
  userId: Number,
  avatarSrc: String,
  post: String
});
onMounted( () => {
  console.log(props.username);
});
</script>
```

The preceding code shows how to access properties by using `defineProps`. This function accepts an object of properties (in our case `username`, `userId`, `avatarSrc` and `post`) and will return a variable that includes all the reactive properties that have been passed when the component was initialized (eg. `<MyComponent username="simone" />`). Next, we introduced another new feature of Vue.js, `onMounted`. This was introduced in the second chapter as part of the Vue lifecycle. `OnMounted` is specifically triggered when the component is fully rendered on the page.

> **defineProps cannot be destructured**
>
> The returned value of `defineProps` cannot be destructured. Destructuring the return object would result in non-reactive values.

Because we have removed the hardcoded strings and changed our component to use properties, we need to do one more step before we can test it in the browser. Just like HTML elements that accept attributes, we need to define our properties when creating a component instance, which in our case was happening in `TheWelcome.vue`.

Let's see how to update our component to include our newly created properties:

```
<template>
  <SocialPost
    username="Username one"
    userId="usernameID1"
    avatarSrc="https://i.pravatar.cc/40"
    post="This is my post"
  ></SocialPost>
</template>
```

Just like a normal HTML element, we are able to pass our properties directly into the HTML tag. The names of the properties used here are the same as the ones defined in the component. They not only have to match word for word, but they are also case sensitive.

Now that our component has been fully updated, we are able to access our application by running the CLI command to start the dev server (npm run dev) and check the browser (http://localhost:5173/).

Our application should not look any different from the previous version of the component. In fact, most of the work that we did was to change how the component behaves behind the scenes, but not how it looks. This section included multiple topics and Vue features. Let's recap what we learned so far:

- How to declare props
- How to use props
- How to use props as plain strings
- How to use props as HTML attributes
- How to use props within the script tag
- How to use our first Vue lifecycle, onMounted

Remember, properties in Vue js are just like HTML attributes. They allow you to make your component dynamic by exposing values that can be used in any shape or form to make your component unique. These properties can be accessed within multiple parts of your component.

In the next section, we are going to learn how to create one or multiple instances of our component and introduce a new concept: reactive data.

> **Your turn**
>
> Continue to expand the footer component you created in the first section but change the values to be dynamic using props.

Learning Vue.js reactive data with Refs and Reactive

In the previous section, we started to make our component dynamic, but that was just the first of two steps required for our component to be fully reusable. It is time to learn about component state, also known as Data (term used in the Option API) or Refs and Reactive (terms used in the composition API). Being able to set private component information, together with the ability to define component properties, will be our toolset for dynamic and flexible components.

Before we jump into data, we need to go back to the previous section and take a look at the component we just created. If you look carefully, the versions of SocialPost.vue look similar, and there seems to be no actual difference between the hardcoded version and the dynamic version.

So, why have we gone through the trouble of making all those changes if nothing changed? Well, the change is there, we have just not used it yet.

Let's think for a minute about our Companion App and try to understand how the `SocialPost` component would be used. When using a real social platform, we would never expect a single hardcoded post on screen; our timeline will eventually display a large number of dynamic posts. In our first version of the component, the one that held hardcoded values, creating the component multiple times would have just resulted in the same post author and title showing over and over again. But with the dynamic version that we have created, we have the chance to pass different values to our props, allowing us to create multiple unique posts. Let's see how this would look in practice by creating a second post:

```
<SocialPost
  username="Username one"
  userId="usernameID1"
  avatarSrc="https://i.pravatar.cc/40"
  post="This is my post"
></SocialPost>
<SocialPost
  username="Username two"
  userId="usernameID2"
  avatarSrc="https://i.pravatar.cc/40"
  post="This is my second post"
></SocialPost>
```

Creating dynamic components is a very powerful tool in web development. It allows us to reuse the same component and simplify our development efforts. Even if making the component dynamic is a step forward from our initial hardcoded example, it is still in need of some improvement. It is time to learn about reactive data and see how it can help us simplify the HTML of our component.

> **Keep the logic away from the HTML**
>
> A clean component is one that has most of its logic encapsulated within the `<script>` tag and has very clean HTML. It may be tempting to add some of the logic into the HTML, but this results in hard-to-maintain components.

Adding Refs or Reactive to our component allows us to remove static data from the `<template>` section of our SFC and allows us to add some dynamicity to our code.

The definition of Refs and Reactive can be a set of primitive data, objects and arrays used by a component instance to define private reactive data (state)."

This notion is not new. In fact, native HTML components hold their own state too. For example, a video component may hold a state of started or stopped, while a dropdown may hold its selected value or an internal state of expanded or collapsed.

In Vue.js and other major frameworks, reactive data is not only used to declare a state (eg. Holding the current state of a sidebar visibility of open or closed) but also to hold component data that is used internally by the component to provide a specific feature.

In our case, we are going to use private data to try and move the information on the individual posts into an array. This will allow us, in future chapters, to use an external tool, such as an API, to fetch this data dynamically.

Before we jump into the actual implementation details, let's define the difference between Refs and Reactive:

- **Refs**: Allows the declaration of primitive values such as string, number, and Boolean and more complex types such as arrays and objects

- **Reactive**: Allows the declaration of objects and arrays and cannot be used for primitive values

	Refs	Reactive
String - "Hello"	✓	✗
Number - 123	✓	✗
Boolean - false	✓	✗
Array - ["Hello"]	✓	✓
Object - { "name": "Bob" }	✓	✓

Figure 3.11: Table of supported and unsupported types for Refs and Reactive

> **Using objects**
>
> Some developers prefer to use Refs for everything, while others like to split the usage depending on the type being assigned. What I am going to show you may be opinionated, and you are free to change your usage to better align with your preferences.

In the course of this book, we are going to use Refs for primitive values such as string, numbers, and Booleans and reactive for arrays and objects.

The main differences between Refs and Reactive are not only in the values that they can hold but also in the way they are used. We are going to make two changes to our component to better understand the difference between Refs and Reactive.

First, we are going to introduce Refs by modifying the `SocialPost` component by adding a new feature to it. Then, we are going to learn about Reactive by moving the post information (`userId`, `avatar`, `name`, and `post`) into an array to simplify our HTML.

Adding Refs to SocialPost.vue

Having the ability to define private data for a component is very powerful. We have seen how a component may need to receive information from its parent by defining properties, but there are times when the component needs to handle its own state. In this section, we are going to make some modifications to our component, `SocialPost.vue`, by providing it with the ability to be selected.

For this feature to be implemented, we need to make three changes to our component:

- We are going to create a private variable called `selected`
- We are going to assign a specific style when the component is selected
- We are going to modify the value of `selected` when the component is clicked

Let's start by creating our first private variable. As mentioned earlier, this will be done using `ref`. This is a method provided by the Vue library that accepts a value that is used to initialize it. For example, if I would like to generate a variable for my name, I would write `const name = ref("Simone")`. In our case, `selected` is going to be a Boolean, and it is going to be initialized with the value `false` because the component is expected not to be selected when first rendered:

```
<script setup >
import { onMounted, ref } from 'vue';
const selected = ref(false);
const props = defineProps({
...
```

As displayed in the preceding code snippets, declaring `ref` is quite simple. First, we import it from Vue, and then we can call it by passing it the initial value of our variable. The rest of the component is omitted but is unchanged from the previous sections.

Next, we create a style for our `selected` state and find a way for this style to be added dynamically when the value of `selected` changes. Let's start by creating a new class called `SocialPost__selected` and adding a white border when this class is active:

```
<style lang="scss">
.SocialPost{
  &__selected{
    border: white solid 1px;
  }
  .header { ...
```

We are going to add our new style to `SocialPost.vue`. Thanks to the help of SCSS, we can define the new class name by using the & helper in `&__selected`. If you have never seen this syntax before, it is a SASS feature that will automatically replace the & with the name of the parent declaration. So, in our case, `__selected` is going to be prepended with `.SocialPost`, creating `.SocialPost__selected`. SASS is not required, and you can achieve these styles using plain CSS, but I have decided to add it to show you the flexibility of Vue with Vite and to help you experience what a real application may utilize.

To make the selected post stand out, we just declare a white border around the component.

Now it is time to assign this class to our component, but we want to do this dynamically depending on the value of `selected`. Our code is going to look like this:

```
<template>
  <div
    class="SocialPost"
    :class="{ SocialPost__selected: selected}"
  >
  <div class="header">
  ...
```

We have just introduced a new feature of Vue.js. In fact, assigning dynamic classes is not possible with plain HTML, but Vue has just the right feature for us.

In *Accessing properties in a Vue.js SFC* , we mentioned that prepending an attribute with `:` allows us to provide it dynamic value, and in the case of the `class` attribute, allows us to assign one or more dynamic classes.

The `:class` attribute accepts an object that is applied to a specific class if its value is truthy. So, in our case, it is going to assign a class of `SocialPost__selected` if the value of `selected` is true.

We are now ready for our last step, the last part of our component enhancement that will allow us to toggle our component and show its selected state.

So far, we have created a specific style and declared a variable that stores our state. What remains is to modify our state variable, `selected`, when the component is clicked. We are going to do this by using the `@click` attribute in the root of our component:

```
<template>
  <div
    class="SocialPost"
    :class="{ SocialPost__selected: selected}"
    @click="selected = !selected"
  >
  <div class="header">...
```

By using the native `@click` event handler and some basic JavaScript, we are able to modify our `selected` variables and update our component state.

If you are not familiar with the syntax used here, by writing `selected = !selected`, we are changing the value of `selected` to be the opposite of the current value. So, if the current value is `true`, it will set it to `false`, and vice versa.

If we run our application and click on one of the components, we should see the following result:

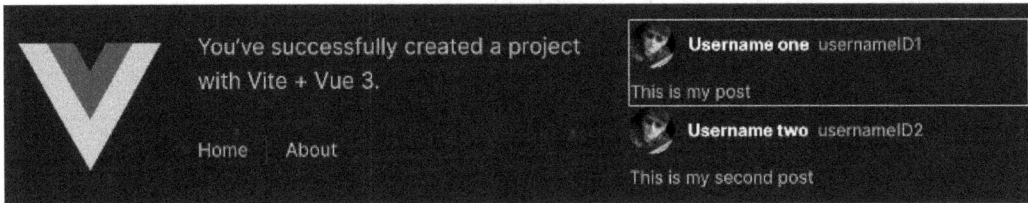

Figure 3.12: Companion App displaying two posts one of which in a selected state with a white border.

We have now learned how to declare and use Refs to define a component state. In the next section, we will move on to our parent component, TheWelcome.vue, and learn how to use **Reactive**.

Using Reactive to host our post information

Keeping clean HTML is the key to a maintainable application, so in this section, we are going to use Reactive to improve TheWelcome.vue. We are going to declare a private variable of the array type. As we already mentioned above, we are going to use Reactive to declare and manage Arrays:

```
<script setup>
  import { reactive } from 'vue';
  import SocialPost from './molecules/SocialPost.vue'
  const posts = reactive([]);
</script>
```

The use of Reactive is very similar to Ref because it needs to be imported from the Vue library and initialized with a base value. In our case, we have assigned our variable a name of posts.

In the preceding code, the value assigned is an empty array, but we need to change this to include the value of the actual posts currently held in the HTML. Our Reactive initialization will be changed to the following:

```
<script setup>
import { reactive } from 'vue';
import SocialPost from './molecules/SocialPost.vue'
const posts = reactive([
  {
    username: "Username one",
    userId: "usernameId1",
    avatarSrc: "https://i.pravatar.cc/40",
    post: "This is my post"
  },
  {
    username: "Username two",
```

```
      userId: "usernameId2",
      avatarSrc: "https://i.pravatar.cc/40",
      post: "This is my second post"
    }
  ]);
</script>
```

Now that our variable is ready, it is time to change the content of HTML to use our Reactive value. Just like properties and Refs, we can use this directly in the HTML.

We will access the information in the first post by using `posts[0].username`, `posts[0].avatar`, and so on. Just like we did previously, we are going to inform Vue.js that the value of our props is dynamic by prepending it with `:`. The component should look like this:

```
<template>
  <SocialPost
    :username="posts[0].username"
    :userId="posts[0].userId"
    :avatarSrc="posts[0].avatar"
    :post="posts[0].post"
  ></SocialPost>
  <SocialPost
    :username="posts[1].username"
    :userId="posts[1].userId"
    :avatarSrc="posts[1].avatar"
    :post="posts[1].post"
  ></SocialPost>
</template>
<script setup>
  import { reactive } from 'vue';
  import SocialPost from './molecules/SocialPost.vue'
  const posts = reactive([
    {
      username: "Username one",
      userId: "usernameId1",
      avatar: "https://i.pravatar.cc/40",
      post: "This is my post"
    },
    {
      username: "Username two",
      userId: "usernameId2",
```

```
    avatar: "https://i.pravatar.cc/40",
    post: "This is my second post"
  }
]);
</script>
```

Our template has now been cleaned from the hardcoded value, which has been replaced with dynamic values declared using Reactive. Defining components using dynamic values with Refs and Reactive will be the foundation of your Vue components for the rest of your career.

Summary

This chapter has introduced you to some basic Vue features, and we have defined the first component for our Companion App. We started the chapter by learning how to create a Vue application using the CLI and looked at its folder structure. We then created our first Vue component. By doing so, we learned how to write and use components with the SFC syntax. We then changed our static component to make use of dynamic properties. Finally, we learned about component state and learned how to use Refs and Reactive data by enhancing the functionality of our components.

> **Your turn**
>
> Use the notion of Ref and Reactive in another component. This could be done in the `footer.vue` file that you previously created by moving the `link` value and `src` into a Reactive property, just like we did for our posts.

In the next chapter, we are going to continue our mission to learn Vue by introducing Vue directives. Directives are Vue-specific attributes that give us the ability to meet complex requirements with simple code. We are going to first introduce the notion of directives and then create new components or update existing ones to learn about the different built-in directives available within the Vue framework.

4

Utilizing Vue's Built-In Directives for Effortless Development

In the previous chapter, we saw how Vue.js features, such as dynamic classes, props, and private state, help us simplify our development. In this chapter, we will continue to build on top of our previous topic by introducing you to the world of Vue.js built-in directives. These directives are the first real taste of Vue.js and will provide you with the tools necessary to make your code more dynamic. The main goal of these directives is to simplify your workflow and make your development easier. Throughout the chapter, we will introduce two new components called `TheHeader.vue` and `SocialPostComments.vue`, alongside continuing to work on existing components such as `SocialPost.vue` to make them more dynamic and ready for future chapters.

We are going to break up the chapter into three different sections:

- Displaying text with `v-text` and `v-html`
- Handling element visibility with `v-if` and `v-show`
- Simplifying your template with `v-for`

By the end of the chapter, you will have a basic understanding of some of the Vue.js built-in directives and be able to toggle component visibility with `v-if` and `v-show`, and will learn how to simplify your HTML templates by using `v-for` and render text or HTML with `v-text` and `v-html`.

Technical requirements

The code required for you to complete the chapter can be found on GitHub within a branch called CH04. To pull this branch, run the following command or use your GUI of choice to support you in this operation:

```
Git switch CH04
```

Do not forget that the repository can be found at https://github.com/PacktPublishing/Vue.js-3-for-Beginners.

There are some small changes to the branch compared to the end of the previous chapter. These changes were added to simplify the book and allow us to focus on the learning material instead of spending too much time on setting things up.

Displaying text with v-text and v-html

HTML is a very powerful tool, but it does have its limitations that forces us to have very verbose and hard-to-maintain code. If you have ever worked on a fully static site without a framework, you may have encountered long HTML pages with many duplicated code blocks. This is where Vue directives come in handy.

Vue's directives are described on Vue School website as *"special HTML attributes that allow us to manipulate the* **Document Object Model (DOM***)."*

In previous chapters, we have seen how Vue uses existing attribute syntax to add functionality such as component properties. Vue directives follow a similar approach that creates new functionality by using syntax that resembles existing HTML.

The preceding description says that Vue.js directives allow us to manipulate the **DOM**, but are there any native HTML attributes that do something similar? The answer is yes.

HTML provides us with attributes such as "value" in a <input> element that assigns the visible value. We also have "rows" and "column" in a <textarea> that define the size of our element. The list could go on for a long time, but I think it is good to illustrate what directives can help us achieve.

Vue's built-in directives, allow us to modify the DOM and enhance the flexibility of a component with the use of the element attribute.

Following are the 15 built-in directives of Vue:

- v-text: Sets the innerText of a given element
- v-html: Sets the innetHtml of a given element
- v-show: Toggles the element visibility (display:hidden)
- v-if: Renders the element if a condition is met

- `v-else`: Displays the element if the previous condition (`v-if` or `v-else-if`) was not met
- `v-else-if`: Displays the element if the previous condition (`v-if`) was not met, and the current one is fulfilled.
- `v-for`: Renders the element or template block multiple times based on the source data
- `v-on`: Triggers a callback when an event is triggered
- `v-bind`: Dynamically binds one or more attributes or a component prop to an expression
- `v-model`: Creates a two-way binding on a form input element or a component
- `v-slot`: Defines a placeholder used by the parent component to be replaced with arbitrary markup
- `v-pre`: Skips compilation for this element and all its children (usually used to display code blocks)
- `v-once`: Renders the element and component once only and skips future updates
- `v-memo`: Memoizes a sub-tree of the template
- `v-cloak`: Used to hide an uncompiled template until it is ready

Some of the preceding directives, including `v-memo`, `v-once`, and `v-cloak`, are for more advanced users and will not be covered during the course of the book, but if you are interested, you can find more information about these in the Vue.js official documentation (`https://vuejs.org/api/built-in-directives.html#v-once`). The rest of the directives will be explained and accompanied by coding examples.

> **Good to know**
>
> All built-in directives start with a prefix of `v-`. This is done to be able to separate the framework attributes from native attributes available in the HTML. This standard is also used to create custom directives that should always start with the same prefix.

Let's start with our first directive, `v-text`.

Applying dynamic text to our component using v-text

Vue.js directives are a very strong addition to Vue.js feature. Not only because they provide extra functionality, but also because they help to write cleaner and more readable code.

When asked about how important code readability is, I always steer people into reading the book *Clean Code* by *Robert Martin* in which he states the following:

> *"Indeed, the ratio of time spent reading versus writing is well over 10 to 1. …*
> *[Therefore,] making it easy to read makes it easier to write."*

Let's start to work with some of these directives and see how they can be plugged in our Companion App.

In this chapter, we are going to work on a new component that will be used as the header of our companion app. This component has already been created in the repository and you can find it within the `components/organisms/` folder.

As you can deduce from the folder in which the component is saved, this component is going to be an "organism" within our atomic design structure that we discussed in *Chapter 1*.

Before we move on, let's explain why our component is going to be called `TheHeader.vue`. The file name of Vue components needs to always be unique and different from native HTML elements, and this prevented us from simply calling it `header` as it would have clashed with the HTML `header` native element. To avoid these issues, Vue.js component names should always be made up of two words written using Pascal case, where the first word and any additional words start with an uppercase letter. There are some cases in which finding two words to make up a component name can be hard. When this case arises, we can prefix the component with the word `the`. Using this approach for our header component will then produce a component name of `TheHeader`.

This component is going to show a logo, a title, and a link to our user profile. If we run our application using the `npm run dev` command, we will see the following:

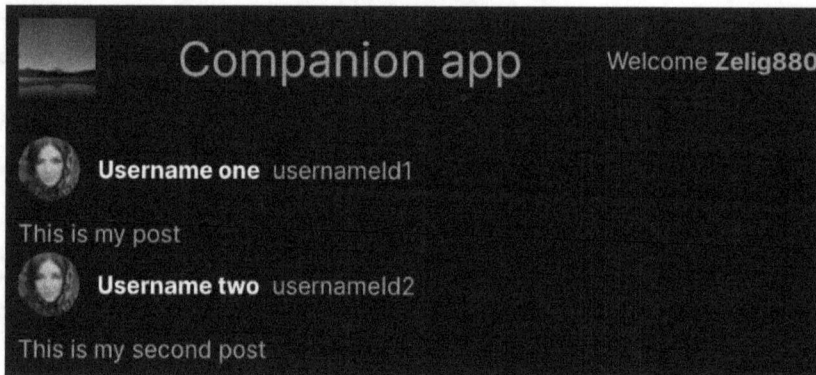

Figure 4.1: Companion app landing page displaying the new header

The code of our component is currently plain HTML with hardcoded data, and it is up to us to update it to make it dynamic.

If you have ever used the internet, you will probably know that a username like the one displayed in the preceding screenshot should not be hardcoded, but rather should be dynamic.

So, our first step is to remove this hardcoded value from the HTML and turn it into a Ref instance, just like we learned in the previous chapter.

First, we are going to create a `Ref` variable with the username in our script tag:

```
<script setup>
import { ref } from 'vue';
```

```
const username = ref("Zelig880")
</script>
```

Then we are going to use our first built-in directives to assign this variable to the HTML element, as shown in the snippets below:

```
<header>
  <img src="https://picsum.photos/50/50" />
  <h1>Companion app</h1>
  <a href="#">
    Welcome
    <span v-text="username"></span>
  </a>
</header>
```

Using the v-text built-in directive will automatically change the innerText of the element with the value of our JavaScript variable username.

Of course, in our example, the name is still hardcoded as the Ref is predefined in our script, but this will later be changed to dynamically be fetched from an API.

If you remember from our previous chapter, we already managed to assign dynamic text to replace the post, name, and username in our SocialPost.vue file. We achieved that using the mustache interpolation syntax {{ }}, just as we did to display our post:

```
<div class="post">{{ post }}</div>
```

So, at this point, you may be asking yourself, what is the actual difference between these two methods?

The answer is not going to be exciting—in fact, both of these methods will compile into the same code at render time. The two methods, even if they achieve the same goal, have different syntax, one defined as an attribute using v-text and the other directly written as part of element content using {{ }}, and will appeal to different developers to align with their coding style.

Another small difference is that v-text will replace the full content of innerHTML while using the mustache interpolation will allow you to replace just part of the text. Let's see an example that illustrates these differences:

```
// v-text
<a href="#">
  Welcome
  <span v-text="username"></span>
</a>
// mustache interpolation
<a href="#">
```

```
    Welcome
    {{ username }}
 </a>
```

In the first example that uses `v-text`, we had to add a `<span>` element to be able to use the directives. In fact, if we had assigned the `v-text` directives directly to the `<a>` tag, the "Welcome" text would have been overridden.

In the second example that uses the double curly braces, we can see that we are able to change just part of the string without the need for extra markup, making it more flexible.

In conclusion, the use of `v-text` or mustache interpolation can be interchanged and there is no right or wrong, it is totally a personal preference.

Modifying the inner HTML of an element with v-html

In the previous subsection, we used our first Vue.js directive, `v-text`. We learned that this directive replaces the `innerText` of a given component, but what if we would like to replace the text with a dynamic string that includes HTML? This is where `v-html` comes in handy. Let's dig deeper into how this can be used and apply it in our application.

> **Rendering HTML can be dangerous**
>
> Rendering HTML dynamically needs to be done carefully as it can be very dangerous aand can lead to XSS vulnerabilities. Only use it for trusted content and never on user-provided content.

Before we explain *how* `v-html` is used, we should first try to understand *why* you would use this technique in your code. In fact, why would you add HTML within a variable when you can write it directly in the `<template>` section of your component?

I asked myself this question at the start of my career, but I then realized that there are a couple of cases in which `v-html` can be useful, including the following:

- **For content coming from a Content Management System (CMS)**: `v-html` can be useful when developing an application that receives its content (such as blog posts) from a CMS. In some cases, this is received as HTML.

- **Third-party plugin output**: There are some third-party plugins such as code visualization and canvas visualization plugins that may require the use of HTML.

- **Hardcoded SVGs or icons**: It is very common for small projects to create a library of hardcoded assets such as icons and SVGs.

As you can see, the real uses of `v-html` are restricted to some very specific scenarios and it is not recommended for all string interpolation usage.

Our header component, `TheHeader.vue`, currently includes a hardcoded SVG to display our logo. We can use `v-html` to make the image dynamic:

1. First, create a variable called `logo` to hold our SVG code:

```
<script setup>
import { ref } from 'vue';
const username = ref("Zelig880");
const logo = `<svg
    height="50"
    width="50"
    viewBox="0 0 210 210"
  >
  <polygon
    points="100,10 40,198 190,78 10,78 160,198"
    style="fill:grey;"/>
  </svg>`;
</script>
```

2. Apply the `v-html` directives in our template:

```
<template>
  <header>
    <div v-html="logo" ></div>
    <h1>Companion app</h1>
    <a href="#">
      Welcome
      <span v-text="username"></span>
    </a>
  </header>
</template>
```

All done! Our application will now display a star as the logo, instead of our previous image, as shown in the following screenshot.

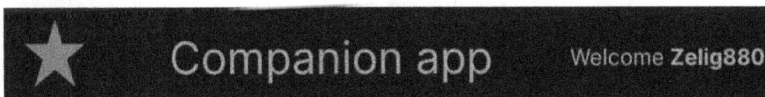

Figure 4.2: Companion app header

This was a simple example that did not really make full use of the dynamic use of `v-html`. In fact, in a case like this, it may be more suitable to just place the SVG directly within the HTML as there is no need to create a dynamic property for something that is static.

It is time to move forward and cover how to toggle the visibility of our component. In the next section, we will learn all about v-show and v-if and start to see how using a framework such as Vue.js can really simplify our code.

Handling element visibility with v-if and v-show

This section is all about element visibility. We are going to learn about two directives that can actually achieve the same result—toggling the element's visibility—but with two underlying differences.

Toggling an element's visibility means the ability to show or hide an element in our rendered HTML. This is a very common feature on the web today. This may be as simple as a dropdown appearing, a modal showing when a button is clicked, or more data being shown when a "show more" link is clicked.

Toggling visibility is the first of many features that support the increased adoption of many frontend frameworks by making the user experience flawless and reactive. First, we are going to add a button in our SocialPost.vue file that will show only if the article has comments, then we are going to display a new component called SocialPostsComments.vue that will be toggled by the new button and will be used in later sections to display a list of posts.

Hiding elements from the DOM with v-show

Having the ability to show or hide an element can be useful for extremely complex situations, but also for very simple cases such as the one that we are about to cover.

If you look at the structure of our posts in TheWelcome.vue, you will notice that the posts object now includes extra information such as comments and tags that are not displayed yet.

Our first step is to display a button in SocialPost.vue that the user will be able to click to show or hide post comments. Because not all posts have comments, we would like to show this button just for articles that include comments. To achieve this, we are going to use v-show.

As with all other directives, v-show is just an element attribute that accepts a value. In this case, the value that is accepted is a **Boolean**. A Boolean is a logical type that can only accept two values: true or false. When v-show receives a Boolean, it will *show* the element if its value is truthy, as in <button v-show="true" />, or *hide* the value if the value is false, as in <button v-show="false" />:

```
<template>
<div

  ...
  <div class="post">{{ post }}</div>
  <button v-show="comments.length > 0">
    Show Comments
  </button>
```

```
    </div>
  </template>
```

In the preceding code, we have accessed our `SocialPost.vue` file and added a new button element. Then since we just want to display this new element only for posts that have comments, we have used our newly introduced `v-show` directive and used the comments properties to define when to toggle the element visibility.

So, in our case, if our post has no comment, the value of `comments.length > 0` will be equal to `false`, and when `v-show` receives a `false` value, it will hide the element. On the other hand, if the post has comments, the value of `comments.length > 0` will be truthy and the button will be shown. The preceding code will display the following result:

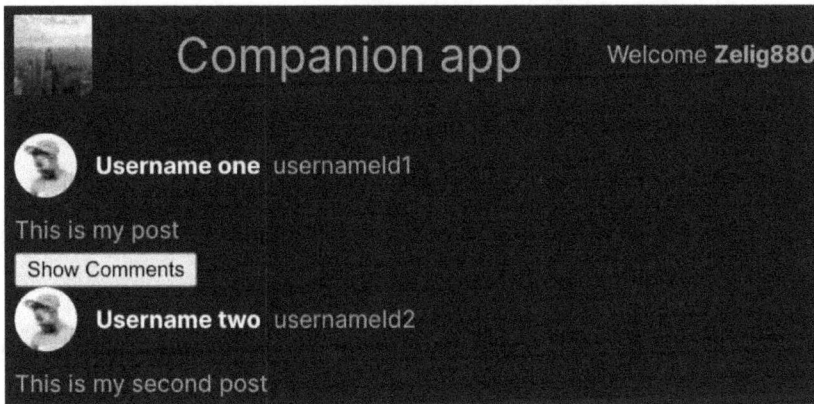

Figure 4.3: Companion app displaying the result of the preceding code snippet
with one of the posts showing the "Show comments" button

Please note that when using `v-show` we do not actually remove the item from the DOM, but just hide it. In fact, if we analyze the DOM using the browser debugger, we can see that the button is available in the second post, but just hidden.

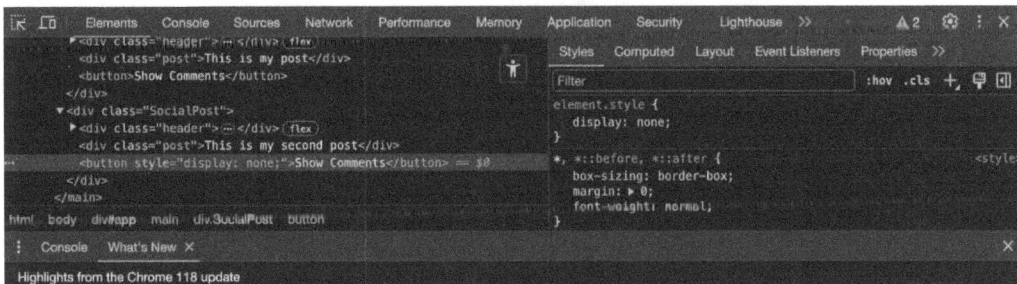

Figure 4.4: Chrome DevTools Elements tab showing the button in
the DOM and its style defined as "display:none"

Even if this seems to be of very little importance for now, it will make more sense after the next section where we will learn about `v-if`.

Keeping the DOM clean and performant with v-if

In the previous subsection, we added a button to show comments, but these comments are still not displayed on the screen. In this section, we are going to add these comments to our UI by working on a new component and then enhance the **Show comments** button to toggle the visibility of these components on and off.

The `v-if` directive works in the same way as `v-show` did. It accepts an argument that will either show or hide the given element depending on it being *truthy* or *falsy*.

In our case, we are going to create a new Ref called `showComments` and this is going to be used to toggle our element visibility. Let's see how to achieve this step by step in our `SocialPost.vue` file.

1. We are going to define our new Ref in the <script> block of the **SFC (Single File Components)**:

    ```
    const showComments = ref(false);
    ```

2. Import the new component at the top of the script tag:

    ```
    import SocialPostComments from './SocialPostComments.vue';
    ```

3. Add that component to our HTML. We do so by passing it a property containing the comments like so:

    ```
    <SocialPostComments :comments="comments"/>
    ```

4. Add the `v-if` directives to make sure that the component is shown only when our showComments private data is true:

    ```
    <SocialPostComments
      v-if="showComments"
      :comments="comments"/>
    ```

5. If we checked the application now, we would notice that the new component is not displayed yet. This is because the value of showComments is set to `false` and there is no way to change it. Let's fix the problem by allowing our "Show Comments" button to change the value of our Ref:

    ```
    <button
      v-show="comments.length > 0"
      @click="showComments = true"
    >Show Comments</button>
    ```

As shown in the preceding code, we used the native `click` event that triggers some basic JavaScript to change the value of our Ref. Thanks to Vue reactivity, our UI will automatically re-render when the value of showComments changes. If you would like to be able to toggle the component on and off, you could write showComment = !showComment to make sure that the value of showComment will be the equal to the *opposite* of the current value of showComment.

The component without its style should look like this:

```
<template>
  <div
  class="SocialPost"
  :class="{ SocialPost__selected: selected}"
  @click="selected = !selected"
>
    <div class="header">
      <img class="avatar" :src="avatarSrc" />
      <div class="name">{{ username }}</div>
      <div class="userId">{{ userId }}</div>
    </div>
    <div class="post">{{ post }}</div>
    <button
      v-show="comments.length > 0"
      @click="showComments = !showComments"
    >Show Comments</button>
    <SocialPostComments
      v-if="showComments"
      :comments="comments"
    />
  </div>
</template>

<script setup >
import { onMounted, ref } from 'vue';
import SocialPostComments from './SocialPostComments.vue';

const showComments = ref(false);
const props = defineProps({
  username: String,
  userId: Number,
  avatarSrc: String,
  post: String,
  comments: Array
});
```

```
onMounted( () => {
  console.log(props.username);
});
</script>
```

Before we move on to the next section, in which we are going to learn how to render a list of items, we need to clarify the difference between v-if and v-show. In fact, until now we have not defined what really makes them different and there seems to be no reason why we could not use v-show in the previous example.

In the previous section, we emphasized the fact that when using v-show, the HTML element is rendered in the DOM but hidden. However, if we look at the same aspect of v-if, we would notice that the element is missing altogether and is replaced by an HTML comment of <!--v-if-->:

```
▼<div id="app" data-v-app> grid
  ▶<header>⋯</header> flex
  ▼<main>
    ▼<div class="SocialPost">
      ▶<div class="header">⋯</div> flex
        <div class="post">This is my post</div>
        <button>Show Comments</button>
        <!--v-if-->
      </div>
```

Figure 4.5: DOM extract of the SocialPosts.vue component

As small as this difference may seem to a new set of eyes, this is actually very important for three main reasons:

- Without the ability to omit elements from HTML, we will encounter errors if we try to render a component (even if hidden) that requires a specific value. For example, in our case, if we had used v-show the Vue compiler would have tried to render the component, but this would have failed as in some cases the comments variable would not have been available.

- Rendering lots of unused elements in the DOM can really affect your performance. Let's say for example that we had 150 comments on each post, and we used v-show. Now the DOM would have hundreds and hundreds of hidden nodes that may never be used. Using v-if allows us to make sure that these nodes are only rendered when needed.

It is time to move forward to our next section where we will be learning a new directive called v-for. This directive allows us to render a list of items, such as our comments, by automatically duplicating our HTML.

Simplifying your template with v-for

The directive covered in this section is called v-for and will be used to render a list of elements. If you have learned any programming language, you have probably been exposed to the concept of a **for loop**. for loops are used in programming languages to iterate through a list one entry at a time, and the v-for directive is no different.

Using v-for allows us to render a specific list by re-rendering the same element (or set of elements) multiple times. Using v-for not only simplifies our **HTML**, but also allows us to render dynamic lists that would be impossible to render ahead of time unless we knew the actual number of entries.

Let's see how the native HTML implementation and a Vue.js implementation of a simple list compare:

Native HTML	Vuejs with v-for

`<ul>`
`  <li>One</li>`
`  <li>Two</li>`
`  <li>Three</li>`
`  <li>Four</li>`
`</li>`
	`<ul>`
	`  <li v-for="item in list">`
	`    {{ item }}`
	`  </li>`
	`</ul>`

Figure 4.6: Comparison of native HTML and Vue.js implementations of a list

As shown, the implementation of a v-for directive is very similar to other for loops that you may have seen before. The directive accepts a parameter in the format of "item in array" just as in major programming languages.

Even if the preceding example does not seem very impressive, this will change over time as v-for is going to become the go-to directive and will help you achieve complex requirements with elegant code.

Let's continue the development of SocialPostComments.vue file and utilize v-for directive to display the different comments for a specific post.

The steps required for us to use a v-for directive are the following:

1. Get access to a variable or property that is an object or array. In our case, it is available as a property.

2. Create the HTML for the first element. This helps to speed up the HTML development.

3. Change the list to use v-for and automatically create duplicate entries.

The first step is already set for us in the `SocialPostComments.vue` file as an array property using `defineProps()`:

```
const props = defineProps({
  comments: Array
})
```

Next, we are going to create the HTML that we want to display. As mentioned previously, we are going to hardcode the first entry of the array by accessing it using the array syntax:

```
<div>
  {{ comments[0] }}
</div>
```

The preceding code in which we hardcoded the first comment, may seem irrelevant, but in more complex development, creating `v-for` this way will save you lots of time. I use this process of rendering just one entry of the `v-for` list to focus on the design and development of the HTML. When the list element if fully designed I then move to the next step.

Finally, it is time to implement our `v-for` directive:

```
<div v-for="comment in comments">
  {{ comment }}
</div>
```

The preceding code will render our `<div>` element as many times as the length of our array.

Did you know?

The body of a `v-for` directive can also be used to load other Vue components. This allows you to abstract complex code into its own components and make your application cleaner.

Before we move to our next chapter, we should point out that `v-for` can accept multiple elements and not just a single element as shown previously. To show this in real life, we are going to add a heading to each of the comments by using the `index` array:

```
<div v-for="(comment, index) in comments">
  <strong>This is comment number {{index}}</strong>
  <p>{{ comments }}</p>
</div>
```

The preceding code shows two more additions.

First, we exposed the index of the array by using the `(comment, index)` syntax. Then we improved the HTML contained in our `v-for` loop by replacing a simple string with a multi-element structure.

Summary

This chapter exposed you to the first Vue directives and started to add some magic to your HTML.

In total, you learned how to accomplish three different use cases. First, we showed you how to apply HTML and text with `v-text` and `v-html`. This is very useful to bind dynamic variables to our component template. Then, we found out how to handle element visibility with `v-if` and `v-show`. We discovered that even if they both achieve the same result visually, they are actually different behind the scenes. `v-show` hides the element using CSS while `v-if` removes it from the DOM altogether. Finally, we finished the chapter with a look at `v-for` and how it can be used to iterate within a list and simplify our HTML.

> **Your turn**
>
> It is your turn to do some practice for this section. There are multiple places to practice directives, but these are some specific suggestions.
>
> 1. Refactor `SocialPost.vue` to use `v-show` to display a post's tags. (You need to add a new property for this.)
>
> 2. Refactor TheWelcome.vue to use v-for to load our posts instead of the hardcoded implementation.
>
> 3. Go back to your footer implementation and replace the hardcoded links using v-text and v-for.

In the next chapter, we return to the logical part of our component and enhance our Refs and props by introducing **computed properties**. The second part of the chapter will introduce **methods** that will hold our component logic and help clean up the HTML of our components further.

5

Leveraging Computed Properties and Methods in Vue.js

In the previous chapter, we learned about **Vue** directives and spent most of our time in the `<template>` section of our component. In this chapter, we are going to move our focus to the `<script>` tag and learn how to ensure that the logic of our component is abstracted.

This chapter aims to introduce you to two features: **methods** and **computed**. These Vue.js features are essential to ensuring that your component logic remains performant and readable. We will go over the components that we created in previous chapters and use the preceding techniques to make them easier to read. Throughout the chapter, we will also continue to add further details to features that we have learned about in previous chapters, such as Refs variables and directives.

This chapter is going to be broken down into the following sections:

- Abstracting functionality using methods
- Creating dynamic data using computed properties
- Learning the difference between methods and computed properties

By the end of this chapter, you will be able to abstract your logic within methods, create reactive variables using computed properties, and most importantly, understand the difference between them.

Technical requirements

To be able to follow this chapter, you should use a branch called CH05. To pull this branch, run the following command or use your GUI software of choice to support you in this operation:

```
git switch CH05
```

The code files for the chapter can be found at https://github.com/PacktPublishing/Vue.js-3-for-Beginners.

Abstracting functionality using methods

Most of the components shown within tutorials and training, such as the ones shared in this book so far, always appear to be easy to read and extremely small, but this is not always the case. Components written in real applications can easily get bloated with logic and become very hard to read. Most of the time, this complexity is the result of multiple iterations and feature changes.

It is very important to try and keep the components as clean as possible. The easiest way to achieve this is by abstracting the logic from the HTML and moving it within <script>, but what can we use to simplify our components?

This is where methods come in handy. Methods are JavaScript functions that can be used within a Vue.js component instance. Methods can be described as helpers that allow you to perform an action within your component.

Methods could be used in a wide variety of cases, from fetching data from an API to validating user input, and will be our go-to Vue.js feature to simplify our component's logic.

Writing methods while using the Composition API syntax is extremely simple as methods are just simple JavaScript functions.

It is time to check out the correct branch from our repository as previously mentioned in the *Technical requirements* section and see methods in action.

We are going to look at the HTML of the SocialPost.vue component within the components/molecules folders and try to find some logic that can be extracted. What we are looking for is any JavaScript code that we have written within the HTML elements of our components that can be turned into a function:

```
<template>
<div
  class="SocialPost"
  :class="{ SocialPost__selected: selected}"
  @click="selected = !selected"
>
  <div class="header">
```

```
    <img class="avatar" :src="avatarSrc" />
    <div class="name">{{ username }}</div>
    <div class="userId">{{ userId }}</div>
  </div>
  <div class="post" v-text="post"></div>
  <button
    v-show="comments.length > 0"
    @click="showComments = !showComments"
  >
    Show Comments
  </button>
  <SocialPostComments
    v-if="showComments"
    :comments="comments"
  />
</div>
</template>
```

When you analyze a component, the best candidates for refactoring into methods are as follows:

- Logic associated with events (click, change)
- Logic that requires a parameter (loop index)
- Logic that performs a side effect

In our preceding code, we have two instances of click events that have some logic associated with them: `@click="selected = !selected"` within the first `<DIV>` and `@click="showComments = !showComments"` in the Show Comments button.

These are great candidates to be refactored as that logic should not live within the DOM (Document Object Model) but be moved out within the `<script>` tag of our component. Refactoring this code is done in two simple steps. First, we will create a method within our `<script>` and then we will replace the logic with our newly created method.

Let's create two functions within our component logic called `onSelectedClick` and `onShow-CommentClick`.

Let's update our code:

```
<script setup >
import { onMounted, ref } from 'vue';
import SocialPostComments from './SocialPostComments.vue';
const selected = ref(false);
const onSelectedClick = () => {
  selected.value = !selected.value;
```

```
}
const showComments = ref(false);
const onShowCommentClick = () => {
  showComments.value = !showComments.value;
}
. . .
</script>
```

> **Prefix your event handler with "on"**
>
> You may have noticed that I have prefixed the event method name with the word "on" and suffixed it with the event name (click). This is good practice as it makes the code more readable and helps you identify methods that are associated with events.

Thanks to the Composition API syntax, we can group our functionalities by adding the methods right below the Ref's initialization, making our code clean and easy to read.

Read and write Ref in the <script> tag

You may have noticed that the code has something different from the logic that we used in the HTML. The selected and showComments Refs have a different syntax as they are followed by .value.

Adding .value to a Ref is a requirement when reading and writing Ref within the <script> tag and does not apply to Reactive variables, which can be accessed with normal variables.

Let's see a visual to help clarify the difference between Ref and Reactive:

	Ref	Reactive
Read in <template>	variableName	variableName
Read in <script>	variableName.value	variableName
Write in <template>	variableName = "foo"	variableName = "foo"
Write in <script>	variableName.value = "foo"	variableName = "foo"

Figure 5.1: Table displaying the differences when reading and writing
Ref and Reactive within the <script> and <template> tags

I know that this may be confusing at first, but by the end of the book, you will have mastered this difference as we will iterate and use this multiple times in the course of our development.

> **Why do Ref need .value**
>
> Vue reactivity is built on top of the proxy object that listens for "get" and "set" events of a variable. This proxy is not available in primitive values such as string, number, and Booleans (the type used with ref). To overcome this, these primitives are transformed into objects with a single property of `.value`.

Now that our methods have been created, it is time to call them. We are going to call these methods from the HTML of our component. To do so, we are going to remove the previous logic that was assigned to the `@click` event and replace it with the methods that we just created.

Our main `<div>` will look like this:

```
<div
  class="SocialPost"
  :class="{ SocialPost__selected: selected}"
  @click="onSelectedClick"
>
```

Our `Show Comments` button will transform into this:

```
<button
  v-show="comments.length > 0"
  @click="onShowCommentClick"
>
  Show Comments
</button>
```

As shown in the preceding code, the logic associated with our click events has been moved into the `<script>` tag.

The change we made may seem very irrelevant and not worth it, but small readability improvements such as this one contribute to a code that is not only easy to read but also easy to change.

Let's modify our component one more time and assume that we have been tasked to trigger a log using `console.log` every time the **Show Comments** button is clicked. If we had not abstracted the logic out of the HTML, achieving this would have been very complicated and verbose. However, luckily for us, the methods are now easily accessible within the `<script>` tag of our component and they can easily be expanded:

```
const showComments = ref(false);
const onShowCommentClick = () => {
  console.log("Showing comments");
  showComments.value = !showComments.value;
}
```

Adding `console.log` could not be easier. We We modified the methods methods using our existing JavaScript knowledge, as shown in the preceding code.

> **Your turn**
>
> Try to create your own methods. My suggestions would be to try and add `console.log` to a message when one of the components mounts (you can use `onMounted`) or continue to practice using Ref and Reactive by creating a method that logs how many posts and comments the application has.

You should now be able to refactor your components by using methods and improve the readability of your code. Breaking down complex code into smaller functions will help us keep our components maintainable. In the next section, we are going to look at a similar feature offered by Vue.js called computed properties.

Creating dynamic data using computed properties

In the previous section, we learned how to simplify our component by abstracting the logic of our click events. We are going to continue on the subject of "refactoring" and introduce a new feature called **computed properties**.

When people first learn about computed properties, they usually compare them to methods due to their similarities, but in reality, this feature is more closely related to props and Ref than methods.

Computed properties can be described as follows:

> *Computed properties enable you to create a dynamic property that can be used to modify, manipulate, and display your component data (refs, reactive, and properties).* - `https://blog.logrocket.com/`

So far, we have used Ref, Reactive, and props to pass and display data within our application, but there are times when the data received may need to be formatted, or when we need to create new data altogether.

When the need for dynamic properties arises, a computed property is the correct feature to use.

Before we start to modify the Companion App, let's introduce a couple of real-life examples to help better understand computed properties and their goal.

- *Scenario 1 – student list*

  Let's assume we receive data from an API that returns all students from a school, but we just want to filter out female students. We can use a computed property to create a filtered array.

- *Scenario 2 – toggle on array length*

  In this case, we have been asked to show a button only if five or more entries exist in our list. We can use computed properties to check the length of our entries and create a new property with a Boolean value.

- *Scenario 3 – concatenate values*

  Imagine an application that has a first name and last name as properties, and you wish to have access to a variable including the full name. Well, we can use a computed property to create this value, which will depend on the first and second names.

- *Scenario 4 – snippets*

  Have you ever encountered a blog in which you see just a small part of the blog post and can click "show more" to show the full article? Well, to achieve this, we can create computer properties that take our blog post and just return a certain number of characters.

If you re-read the preceding example, you may notice that they all have something in common. In fact, all the scenarios that we proposed have the following similarities:

- They all create a new piece of data / value that is required within the component

- They do not produce any side effect

- They all depend on another piece of data (props, Refs, or Reactive)

The preceding bullets are present in all of the scenarios that I proposed, not because of my choice, but because they are the prerequisites of computed properties.

A computed property is simply a feature that allows us to use one or more existing pieces of data (Refs, Reactive, and props) to create a new value.

It is time to start and look at some examples to be able to learn how and when to use this feature.

The syntax of a computed property is the following:

```
import { computed } from 'vue'
const test = computed( ... );
```

When using a computed property, we need to first import it from Vue and then assign it to a new constant. The computed property accesses a callback that is going to include the logic required to create a new variable.

Let's see a simple example and show how a computed property that creates a full name would look like:

```
<script setup>
import { ref, computed } from 'vue'
const name = ref("Simone");
const surname = ref("Cuomo");
const fullName = computed( () => {
  return `${name.value} ${surname.value}`;
} );
</script>
```

In the preceding code, we have created a new dynamic property called `fullName`. This will act just like a normal property and can be used everywhere without our components.

Just like we said before, the computed property fulfills our requirements, which are as follows: it creates new data (`fullName`), depends on another value (`name` and `surname`), and returns a value.

Adding computed properties to the companion application

It is now time to update our component and see how we can leverage computed properties to simplify our code even further.

Just like in the previous section, we are going to continue to modify the `SocialPost.vue` file. Let's review the file and try to see what a good candidate would be for a computed property:

```
<template>
  <div
    class="SocialPost"
    :class="{ SocialPost__selected: selected}"
    @click="onSelectedClick"
  >
    <div class="header">
    <img class="avatar" :src="avatarSrc" />
    <div class="name">{{ username }}</div>
    <div class="userId">{{ userId }}</div>
    </div>
    <div class="post" v-text="post"></div>
    <button
      v-show="comments.length > 0"
      @click="onShowCommentClick"
    >
      Show Comments
    </button>
    <SocialPostComments
      v-if="showComments"
      :comments="comments"
    />
  </div>
</template>
```

From the preceding code, we can see that the logic associated with the `v-show` directive has been highlighted. This is our best candidate to be turned into a computed property.

Computed properties need to return a value that depends on another value and has no side effect, and the preceding highlighted logic does just that. In fact, it returns a boolean of `true` or `false`, it depends on the value of `comments`, and it does nothing else that can be defined as a side effect.

Just like methods, to be able to convert this logic into a computed property, we need to move its logic into the `<script>` tag of our component. Let's see the steps required to achieve this:

1. First, we need to import `computed` at the top of our `<script>` tag:

    ```
    import { onMounted, ref, computed } from 'vue';
    ```

2. Then, we create a function that uses a computed property:

    ```
    const hasComments = computed( );
    ```

3. Next, we are going to add our logic as the first argument of the computed method as a callback:

    ```
    const hasComments = computed(() => {
      return props.comments.length > 0;
    });
    ```

4. Last, we replace the existing logic in the HTML with the new computed property:

    ```
    <button
      v-show="hasComments"
      @click="onShowCommentClick"
    >
      Show Comments
    </button>
    ```

With the preceding code, we have now created a new property within our component called `hasComments`. This property, just like every other Vue.js variable, is responsive and will change as soon as the `comments` array changes.

Increasing performance with cached values

Creating computed properties not only improves the readability of our component but also improves the performance of our application by caching the value.

What this means is that the actual function included within the computed property just gets run once, when the component mounts, but it does not run again when the component renders unless the dependent value changes.

This may not bring a massive improvement to our simple example, but it does make a big difference in large applications where the computed property could actually be a large array with 100s of entries!

Formatting your data

Computed properties are not very common among other major frameworks and they are seen as a new technique/feature for most Vue.js developers. Due to their unique nature, it can be hard at times to grasp and use them in real life. To ensure that the topic has been fully understood, we are going to create a couple of extra examples that make use of computed properties.

If you open the file called `TheWelcome.vue` and check the `posts` Reactive value, you will notice that there are extra parameters in the object. In fact, the value of `likes` and `retweets` has been added to our posts.

```
const posts = reactive([
  { username: "Username one",
    userId: "usernameId1",
    avatar: "https://i.pravatar.cc/40",
    post: "This is my post",
    comments: [
      "great post",
      "amazing post"
    ],
    likes: 2,
    retweets: 1,
    tags: [
      "tag 1"
    ]
  },
```

Figure 5.2: The posts Reactive property including extra likes and retweets entries

In this section, we are going to create a new dynamic property that will hold the total number of interactions. We are going to do so by adding all the different interactions, including comments, likes, and retweets.

This new computed property will be called `interactions`, and in the case in which our post has 2 comments, 2 likes, and 1 retweet, it will return a value of 5 (2 + 2 + 1).

This example should help you understand that the computed property is not just a way to prettify your component, but is actually a powerful tool that helps you enhance the features of your application.

> **Give it a try**
>
> Why don't you give it a try and attempt to create the computed property by yourself? Creating this computed property will require you to put into practice everything you learned until now, so why not take a little challenge?

Successfully creating a computed property to display our interaction requires a couple of steps. The following diagram shows how the data flows through the component, and it will help contextualize the steps we will have to take to fully define our computed property.

Data is set in the parent component

↓

The new properties are added to the instance
of the child component

↓

The new properties are "read" by the child component
using defineProps

↓

Create a new computed that uses the new properties

↓

The computed value is added in the <template> section

↓

The HTML is styled

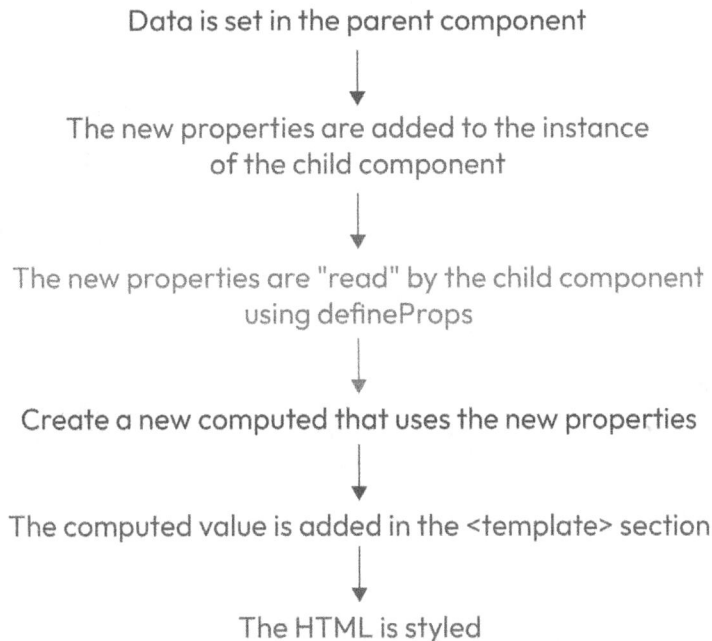

Figure 5.3: Data flow from the parent component to the computed property

Throughout this book, we have already covered each of the preceding entries, but this will be the first time that we are using them all together in one single exercise. Generating a full data flow like this will be very common in your development life, so it is beneficial to get some early practice.

Let's go over the preceding data flow one step at a time:

1.   **Set data in the parent component**

     Let's open TheWelcome.vue and check the parameters available in the posts Reactive within the <script> tag:

```
{ username: "Username one",
  userId: "usernameId1",
  avatar: "https://i.pravatar.cc/40",
  post: "This is my post",
  comments: [
    "great post",
    "amazing post"
  ],
  likes: 2,
  retweets: 1,
  tags: [
    "tag 1"
```

```
    ]
  }
```

As shown in the preceding code, the properties of likes and retweets are already set for us.

2. **The new properties are added to the instance of the child component**:

As explained in a previous chapter, a component requires properties to be passed to it, before it can use them. So, in our case, we need to add the new likes and retweets properties to the instances of <SocialPost>:

```
<SocialPost
:username="posts[0].username"
:userId="posts[0].userId"
:avatarSrc="posts[0].avatar"
:post="posts[0].post"
:comments="posts[0].comments"
:likes="posts[0].likes"
:retweets="posts[0].retweets"
></SocialPost>
```

The preceding example shows how you can add it to the first entry and how you can replicate the same for the second one.

3. **The new properties are read by the child component using definedProps**:

To achieve this step, we need to open the SocialPost.vue file and add the new properties of likes and retweets in the list of definedProps. These are both going to be Number types:

```
const props = defineProps({
  username: String,
  userId: Number,
  avatarSrc: String,
  post: String,
  comments: Array,
  likes: Number,
  retweets: Number,
});
```

4. **Create a new computed property that uses the new properties**:

The computed function is going to use the newly created properties and create a new dynamic property called interactions:

```
const interactions = computed( ()=> {
  const comments = props.comments.length;
  return comments + props.likes + props.retweets;
});
```

Just like every computed property, our preceding function creates a new `interactions` value using the `computed()` function and returns a value, which, in our case, is the sum of different properties.

5. The computed value is added to `<template>`:

```
<div class="interactions">Interactions: {{ interactions }}</div>
```

Here, we used our knowledge of string interpolation to print the value of interactions within the component.

6. **Style the HTML**:

The last step requires us to style the HTML. This can be done by using plain CSS within the `<style>` tag of our component:

```
.interactions {
    font-weight: bold;
    margin-top: 8px;
}
```

After following the steps defined previously, our new property for `interactions` is fully set up and available within our companion application, as shown in the following screenshot.

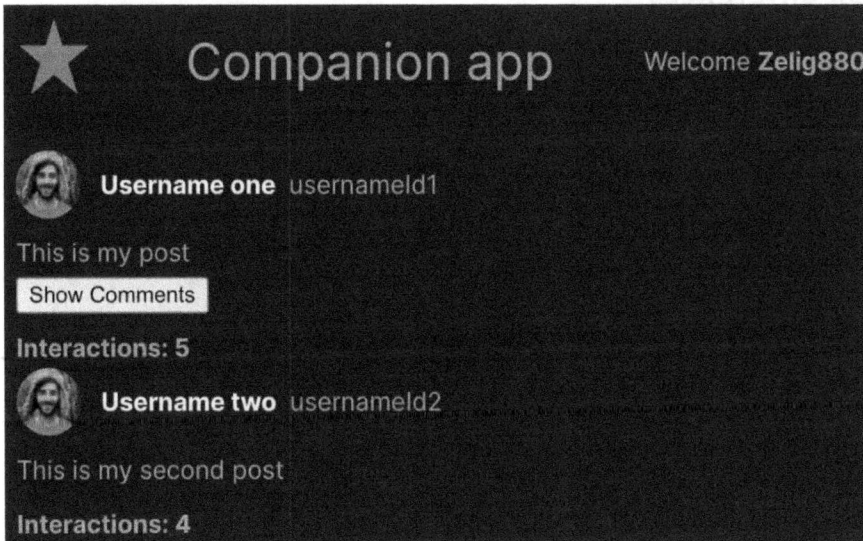

Figure 5.4: Companion application screenshot showing interactions

Computed properties will be a vital part of your Vue.js application and it is vital to understand when you should use them and how to use them.

> **Your turn**
>
> Try to create your own computed property. Open up the `TheHeader.vue` file and try to replace the current implementation of `Welcome {{ username }}` with a new computed property called `welcomeMessage`.

In this section, we learned how to enhance our components by creating dynamic properties using computed properties. We have explained the three factors that make a computed property, which are as follows: it produces no side effect, creates a new variable, and depends on other component variables. Finally, we went over two different coding exercises to better understand the meaning and usage of computed properties.

Learning the difference between methods and computed properties

During my career, I have seen many people misusing methods and computed properties. In this section, we are going to clarify the difference between the two features and provide you with a guide that you can revisit during the course of your early career.

Technical differences

It may be a good starting point to re-iterate what we discussed in the previous two sections and see side by side what the technical differences are between these two Vue.js features:

Ref	Computed
Are called when needed	Are initialised immediately
Can have parameters passed in	Shoud not have parameters
Re-evaluate on every call	Cached based on its dependencies
Does not have to return a value	Needs to return a value
Can produce side effect	Cannot include side effect

Figure 5.5: A table showing the technical differences between methods and computed properties

Let's focus on some of the main differences displayed in *Figure 5.5*:

- **Time of initialization**: Methods can be initialized on demand by the user, while computed properties are triggered during the component creation life cycle.

- **Parameters and dependencies**: While methods can accept parameters, computed properties can just use other component props and data that are called dependencies.

- **Evaluation**: Methods are evaluated every time they are called, making them very expensive, while computed properties are cached and just re-evaluated if any of their dependencies are updated.

- **Side effect**: Side effects occur when a method or a function modifies a value or triggers an action outside of its local scope. So, for example, triggering an API call is a side effect, or changing a variable that is not returned by the method is a side effect.

The goal of a computed property is to create a new variable and it should never include any side effect, while the main scope of methods is to produce side effects.

If someone were asked to define the difference between computed properties and methods in a single sentence, I would say the following:

Methods are helper functions that allow you to complete an action, while computed properties allow you to create a new component property.

How to spot them in your component

In the previous section, we spoke about the technical differences between methods and computed properties, while in this section we will see where we would find these helpers within a component.

Knowing where computed properties and methods are used within the component structure will help you make the correct choice. To better see whether we can make the correct choice, we are going to use a dummy component. This component has a few placeholders delimited by ???.

Spend a couple of minutes to try and understand whether this is going to be methods or computed properties and why you have made this choice. When you are ready, you can read my explanation and see whether it matches your choices:

```
<script setup>
import { reactive } from 'vue'
const post = reactive({
  title: '',
  content: '',
showOnlySnippets: true
});
const fetchBlog = //???;
const onShowAll = //???;
const snippets = //???;
onMounted(() => {
  fetchBlog();
} );
</script>
<template>
  <h1>{{ post.title }}</h1>
  <p v-if="post.showOnlySnippets">{{ snippets }}</p>
```

```
    <p v-else>{{ post.content }}</p>
    <button @click="onShowAll" >Show All</button>
  </template>
```

The first placeholder is fetchBlog. This is going to be a method. The decision is based on the fact that fetchBlog is not only going to trigger a side effect, contacting an external API, but it is also called in the onMounted life cycle. As we previously defined, only methods can be triggered on demand.

The second one is onShowAll. This is going to run an action, something that will be triggered when the user clicks a button. Again, in this instance, we are going to define this as a method because it is triggered by an event, will include a side effect, and will accept parameters.

Lastly, we have snippets. This variable is used in the template just like a variable. As we should know by now, creating dynamic variables is the perfect match for a computed property. In this instance, snippets is going to be a computed property that will depend on the blog content.

Let's see what the updates script may look like:

```
<script setup>
import { reactive } from 'vue'
const post = reactive({
  title: '',
  content: '',
  showOnlySnippets: true
});
const fetchBlog = () => {
  fetch('fakeApi.com').then( result => {
    post.title = result.title;
    post.content = result.content;
  } );
};
const onShowAll = (event) => {
  post.showOnlySnippets = false;
};
const snippets = computed( () => {
  return post.content.slice(0, 100);
} );

onMounted(() => {
  fetchBlog();
} );
</script>
```

With all the examples we have covered so far, you should have obtained a good understanding of these two different features.

Summary

In this chapter, we have introduced methods and used them to clean up the components of our Companion App. We learned the difference between Refs and Reactive and learned how to use them within the `<script>` section of components, and then we moved forward and learned how to create dynamic properties using computed properties. To complete the chapter, we fortified our knowledge of these two topics by going over a few more examples.

In the next chapter, we are going to learn how to create and manage events using Vue event handling. Up until now, we have focused on the individual components, but with the introduction of events and events handling, we will be able to define two-way communication between components. Event propagation is an essential skill to have to be able to develop scalable and clear component-based applications.

6

Event and Data Handling in Vue.js

Major frameworks have gained their popularity not only from their ability to break down big pages into small reusable components but also due to the simplicity with which these components can communicate with each other.

In this chapter, we are going to focus on how the data flows between different components. This exchange of information between components is handled in two different ways: parent to child with properties and child to parent with events.

We are going to start this chapter by introducing a few changes that have been committed in the Companion App since our last chapter. These modifications can be used as a good guide to help you develop your skills and consolidate what you have already learned. We are going to revisit properties by learning about more advanced techniques such as `validator` and `required` by creating our first reusable component in the form of a simple button. Then, we will learn how to handle native events such as `click` and `change`, and finally, introduce custom events that will allow our component to broadcast messages to the rest of the app.

This chapter includes the following sections:

- Exploring the Companion App changes

- Deepening our knowledge of props

- Handling native events in Vue.js

- Connecting components with custom events

By the end of the chapter, you will have learned how to handle communication between multiple components. With this new knowledge, you should be able to move away from the individual components and start to think at a macro level by considering more complex application structures that include multiple components.

Technical requirements

To be able to follow this chapter, you should use a branch called CH06. To pull this branch, run the following command or use your GUI of choice to support you in this operation:

```
git switch CH06
```

This branch is going to include a few changes. These will be explained in the first section of this chapter, but you can start and run the app to get familiar with its updated look and browse the repository.

The code files for the chapter can be found at https://github.com/PacktPublishing/Vue.js-3-for-Beginners.

Exploring the Companion App changes

Until now, the Companion App had a very basic look and feel and its changes were completely made by the code that we wrote within the chapters, but things have changed.

Let's see how the application looks now:

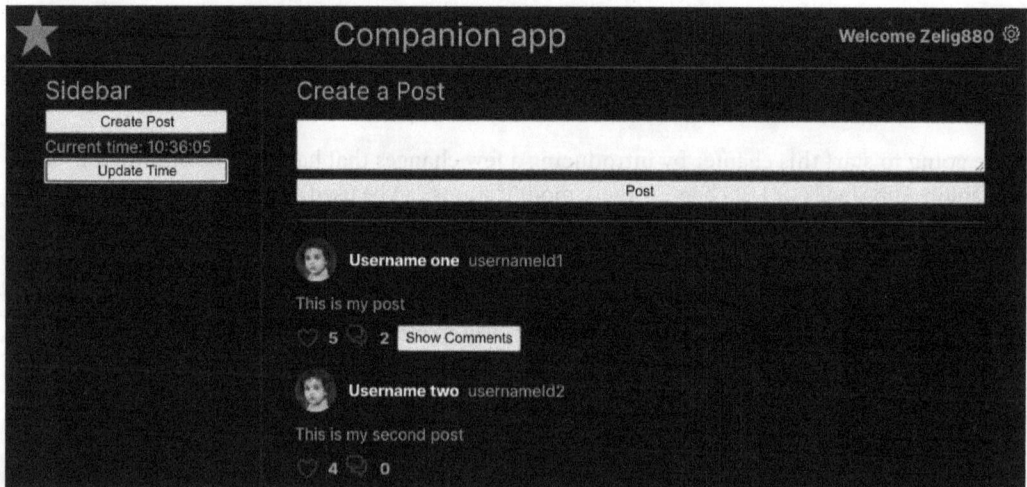

Figure 6.1: Companion app with updated design and components

As you can observe in the previous screenshot, the Companion App has not only received a facelift with some updated styles but it also exhibits new components that will be used in the course of this and future chapters.

These changes have been applied to be able to make the most of the book content, allowing us to focus on the new learning material without having to spend too much time creating the scaffolding and basic structure of components, but you should spend some time exploring the changes made and try to understand why and how things were implemented.

All the changes and modifications that have been added since last chapter, use features that you have already been introduced to, and going through each change is a perfect exercise to enforce your learning.

Two main groups of changes have been made to the app:

- Folder and file changes
- Logic changes

Let's look at these changes in detail, starting with the changes that affected folders and files.

Folders and file changes

In the CH06 branch, a few changes have taken place within the folders and files of the application. The changes were made to start to structure the app like a real production app, moving away from the simple "proof of concept" look and structure that the app had until now.

The changes made in the file structure are as follows:

- Our first atom component with TheLogo.vue was added.

    In the previous chapter, the SVG element of the logo was hardcoded in the header of our Companion App. This has now been moved into its own file and it has been imported into TheHeader.vue.

- A sidebar called SideBar.vue was added to the UI.

    The layout of our app has been modified and we have added a new sidebar in the organism folder with the name SideBar.vue.

- The ability to create posts with CreatePost.vue was scaffolded.

    The main container of our application now includes a new component aimed at creating new posts. This component just includes the HTML structure and some style, but it has no logic yet. The component can be found at CreatePost.vue in the molecules folder.

- We renamed TheWelcome.vue to SocialPosts.vue.

    To better align our components with the newly updated layout, there was a need to remove the very generic TheWelcome.vue component and align it with the component structure. To do so, we have renamed the component SocialPosts.vue as it includes a list of social posts, and also moved it within the molecules folder.

Now that we have found out about all the files and folder changes, it is time to see any code logic changes that may have been included in this update.

Logic changes

Just like the folder changes, the app has gone through a few modifications. These changes will align the app for future changes, and they also give you a glimpse of what it is like to work on a real project by adding more files and a more complex folder structure.

If you were to finish this book with a repository that has a handful of components, you would not experience what it is like to work on a real application, and you will therefore fail to learn how to navigate the code base and, more importantly, structure your code base. For this reason, the app has been enhanced to have some more structures:

- Defining `homeView.vue`: At this moment in time, the application is a single page, but this will change later down the line. To start and align with those changes, we have started to define the **Home** layout (`src/views/HomeViews.vue`) to include the sidebar, header, and main page body. This will allow us to create other pages in the future without the need to duplicate this structure.

- Cleaning up the code in the header: The header file has been cleaned up. We removed the hardcoded logo and added a new icon next to the "Welcome" message. To complete the changes, we also added some styles to our components. Look at this file to remind yourself how to load external components.

- Making `SocialPost` render dynamic content with `v-for`: Up until now, `SocialPost` posts were individually and manually loaded in `SocialPosts.vue` (previously known as `TheWelcome.vue`). The rendering of multiple posts is now dynamic and it is handled by `v-for`. Using `v-for` has not only improved the readability of the component but also made it dynamic, allowing us to add `post` to our list without the need to manually change the HTML.

```
 1  <template>                                          1  <template>
 2      <SocialPost                                      2      <SocialPost
 3-         :username="posts[0].username"                3+         v-for="post in posts"
 4-         :userId="posts[0].userId"                    4+         :username="post.username"
 5-         :avatarSrc="posts[0].avatar"                 5+         :userId="post.userId"
 6-         :post="posts[0].post"                        6+         :avatarSrc="post.avatar"
 7-         :comments="posts[0].comments"                7+         :post="post.post"
 8-         :likes="posts[0].likes"                      8+         :comments="post.comments"
 9-         :retweets="posts[0].retweets"                9+         :likes="post.likes"
10-     ></SocialPost>                                  10+         :retweets="post.retweets"
11-     <SocialPost                                     11+         :key="post.userId"
12-         :username="posts[1].username"
13-         :userId="posts[1].userId"
14-         :avatarSrc="posts[1].avatar"
15-         :post="posts[1].post"
16-         :comments="posts[1].comments"
17-         :likes="posts[1].likes"
18-         :retweets="posts[1].retweets"
19-     ></SocialPost>                                  12      ></SocialPost>
20  </template>                                         13  </template>
```

Figure 6.2: Git differences between the hardcoded version of SocialPost and the one using v-if

We have now finished explaining both the structural and logical changes included in this chapter. Before we move forward, there is one small detail that I want to mention and it is about the `:key` property.

If you carefully check the previous screen, you will notice that the new version of the code (the right-hand side) has an extra property called :key. This property is a requirement when using v-for with Vue components.

The key value is used by the framework to prevent unnecessary re-rendering of the full list. In fact, Vue.js uses this attribute to keep track of which specific component instance changed and just update a specific node instead of the full list.

So, going forward, every time you use the v-for directives, you should remember to also set a unique key using :key. This is usually raised as a warning by linting tools. Providing this value helps Vue.js identify all the different nodes that it creates with v-for, and speeds up re-rendering if any of it changes. You should try to define a key whenever possible, unless the DOM created is very simple.

> **Avoid using an array index**
>
> It is very common to see the array index being used as the :key value for a v-for loop. Unfortunately, this is a very bad practice as the index will change if an array item is removed, forcing Vue.js to re-render the full list and possibly produce bugs that are hard to spot.

We are now up to speed with all the changes we have made. As mentioned previously, you should spend a few minutes familiarizing yourself with the changes and understand what was done and why.

In the next section, we are going to start our journey with data flows and look closely at properties. This subject was already introduced in the previous chapter but it has more to offer, and it is time to learn about it.

Deepening our knowledge of props

In *Chapter 3*, we introduced and started to use properties as a way to pass information from a component to its children. As our understanding of Vue.js is widening, it is time to expand our knowledge of this basic feature.

Just as a quick recap, so far, we have learned that properties (props) are defined using the defineProps compiler macro, like so:

```
const props = defineProps({
  name: String
});
```

Doing so will allow our component to accept a prop called name of type String:

```
<myComponent :name="myName" />
```

In this section, we are going to learn what other configurations properties have to offer. The ability to define a props type is not the only configuration available.

Props configurations

In most cases, just configuring a prop type, as previously shown, is more than enough, but there are times when fine control is needed, and the following configuration will help you.

The syntax we used to declare properties has a name followed by the expected type, `PropsName : Type`, but to be able to use advanced configuration, we need to change the props to accept an object. So, the example that we provided before of `name : String` in object syntax will be as follows:

```
const props = defineProps({
  name: {
    type: String
  }
});
```

Now that the properties have an object, we can add extra configuration.

Multiple types

A Vue.js prop can accept multiple types. To do so, we can just define the types as an array:

```
name: {
  type: [ String, Number ]
}
```

Setting required properties

There are two types of properties, `required` or `optional`. Out of the box, all properties are set to be `optional` by the framework and this is done on purpose to align with the way native HTML handles attributes (e.g., `disabled` or `readonly`).

When developing components, you should always ask yourself whether the component can be rendered without the value of a prop value, and if not, make sure you set the prop as `required`:

```
name: {
  type: String,
  required: true
}
```

When a prop is required, if a user tries to implement the component without it, the component will throw an error and, in some cases, not render the component at all, depending on your settings.

Setting a properties fallback with a default value

When a prop is set to `optional` and it is not passed, the component will give it an `undefined` value. This is okay for most cases, but there are situations in which you would like the properties to have a fallback value.

It is common for beginner developers to achieve this using v-if in the HTML:

```
<button>{{ welcomeMsg ? welcomeMsg : "Welcome" }}</button>
```

This solution is very verbose and should be replaced with a default prop value:

```
welcomeMsg: {
  type: String,
  default: "Welcome"
}
```

When the welcomeMsg prop is not passed, the component will render the Welcome string. Our HTML can now be clean for logic:

```
<button>{{ welcomeMsg }}</button>
```

> **Arrays and object default initialization**
>
> Objects and arrays have a different syntax for default and must be returned by a factory function: default: () => [] or default: () => {}.

Validating your properties

The last configuration that we are going to introduce is the ability to validate the value received. This is extremely helpful when your prop can just accept a specific set of values or if they have to be formatted in a specific way. A validator is a function that receives an argument equal to the prop value and is expected to return true to mark the validation or false to invalidate the validation.

So, for example, if we want to create a prop that can just accept two strings, light or dark, we would set the validator function as follows:

```
theme: {
  type: String,
  validator: (value ) => ["light", "dark"].includes(value)
}
```

With the preceding lines of code, our component props, theme, will just accept these two values. If the wrong value is passed, the component will not render and will throw an error.

Creating a basic button

To put everything that we have learned into practice, we are going to create a simple base button. Base components, such as the one offered by component libraries, accept multiple properties and are usually the best fit to utilize advanced properties settings.

Our button is going to have the following features:

- It will *require* a value that can be a string or number

- If will have an `optional` prop of `width` that defaults to `100px`

- It will have an `optional` theme that only accepts a `light` or `dark` value

The new button can be found in the `atoms` folder under the name `TheButton.vue`:

```
<template>
  <button
    :class="theme"
  >
    {{ value }}
  </button>
</template>
<script setup>
  defineProps({
})
</script>
<style scoped>
button {
  width: v-bind(width);
}
.light {
  background-color: #1DA1F2;
  color: white;
}
.dark {
  background-color: black;
  color: #1DA1F2;
}
</style>
```

The file is mostly defined. Its HTML includes a `button` element with placeholders for our properties and styles ready to accommodate our prop value. All that is left for this component is to define its properties. Before reading the following solution, try and use the preceding information to define the props by yourself.

When fully defined, the defined props should look like this:

```
defineProps({
  value: {
    type: [String, Number],
    required: true
```

```
    },
    width: {
      type: String,
      default: "100px"
    },
    theme: {
      type: String,
      default: "light",
      validator: (value) => ["light", "dark"].includes(value)
    }
  })
```

Prop configurations are extremely powerful and so simple to use. Using them correctly can help save many lines of code in your template and will help you make your component more robust.

In this section, we have learned how to use prop configurations and created a base button to help us understand its real usage. We learned how to provide multiple types, how to set a prop as `required`, how to define a default value, and last but not least, how to validate it.

In the next section, we will start to move our attention to another part of data handling: events. Events are used by children components to communicate with their parent component. We are going to first introduce the native element and then move on to define custom elements.

Handling native events in Vue.js

Since JavaScript's inception, events have always played a vital role in the success of this programming language. For this reason, all JavaScript frameworks have made sure they offer a strong solution to handle native and custom events.

We refer to native events as the ones built in HTML elements and APIs such as a `click` event triggered by `<button>`, a `change` event triggered by `<select>`, or a `load` event triggered by `<img>`.

Just like with props and directives, Vue.js event handling seamlessly merges with the existing native syntax to handle events offered by HTML elements. In native HTML, all events handlers are prefixed with the word on, so a `click` event becomes `onclick` and a `change` event becomes `onchange`. Vue.js follows this convention by creating a directive (which, as we know, starts with `v-`) called `v-on`. So a `click` event in Vue is handled using `v-on:click` and a `change` event is listened to by using `v-on:change`.

You have probably noticed that this is not the syntax that we used in the code base so far. In fact, if you open up `SocialPost.vue`, you will notice that we used a different syntax where the event is prefixed with the @ symbol.

This is just a nice shorthand provided by the Vue.js framework. Using the @ sign not only simplifies the actual writing of the events but is also clearly marked as different from other directives.

Let's see how these different methods would look when applied to `<button>`:

```
// HTML Native
<button onclick="method()">Click Me</button>
// Vue.js
<button v-on:click="method">Click Me</button>
<button @click="method">Click Me</button>
```

Even if the syntax is not completely the same, it is very similar and intuitive to use.

There is a small difference to notice. In fact, in the native method to handle events, the method that is passed to `onclick` is actually called `onclick="method()"`, while in the Vue.js case, the methods are not called and they are just passed as a reference to the method name that will later be called by the framework engine, `@click="method`.

It is time to make some changes to the code base by opening up `SideBar.vue`:

```
<template>
<aside>
  <h2>Sidebar</h2>
  <button>Create Post</button>
  <div>
    Current time: {{currentTime}}
  </div>
  <button>Update Time</button>
</aside>
</template>
<script setup>
import { ref } from 'vue';
const currentTime = ref(new Date().toLocaleTimeString());
</script>
```

This file is the one responsible for the new sidebar being displayed in our layout. This component includes a couple of elements and buttons that will be used as learning material in the course of this and future chapters.

For this section, we are going to focus on the current time displayed on the screen. More specifically, we are going to create the ability to update this time at the click of a button.

Handling events is achieved in two steps. First, we create a method that includes all the logic to trigger the event, and then we attach this to the HTML.

Let's create the method first:

```
<template>
<aside>
  <h2>Sidebar</h2>
```

```
  <button>Create Post</button>
  <div>
    Current time: {{currentTime}}
  </div>
  <button @click="onUpdateTimeClick">Update Time</button>
</aside>
</template>
<script setup>
import { ref } from 'vue';
const currentTime = ref(new Date().toLocaleTimeString());
const onUpdateTimeClick = () => {
  currentTime.value = new Date().toLocaleTimeString();
};
</script>
```

With the preceding change, clicking the button labeled **Update Time** will update the current time shown on the screen. Let's break down how we achieved this.

First, we create a method for our event logic. This method is going to be prefixed with the word on followed by an identifier of the actual event – for example, updateTimeClick – generating a name equal to onUpdateTimeClick.

The logic of this particular event is quite simple, but as we learned in the previous chapter, methods can be complex and also have side effects, so there is no limit to what an event can achieve.

Next, we add the click directive in the HTML. We can either use v-on:click or @click. This will automatically link our Vue method with the native click event.

This example should not have been too different from the one that we already covered in the earlier chapters, but event handling does not stop here. In fact, there are two important concepts that we need to learn about: event modifiers, which we are going to introduce right now, and event parameters, which will be covered later in the chapter.

Event modifiers

While working with events in JavaScript, it is very common to modify the event by either preventing the default behavior (event.preventDefault()) or stopping the propagation of the event (event.stopPropagation()).

This is where the event modifier comes into place. Vue.js offers us a simple syntax to trigger these logics directly from the HTML without the need to navigate through the event object.

Vue.js offers many modifiers – from event modifiers (such as .prevent, .stop, and .once) to keyboard modifiers that allow you to listen to a specific button (such as .enter, .tab, and .space) and, finally, to mouse modifiers that allow you to trigger events on a specific mouse press with .left, .right, and .middle.

Let's break down the syntax of a Vue directive:

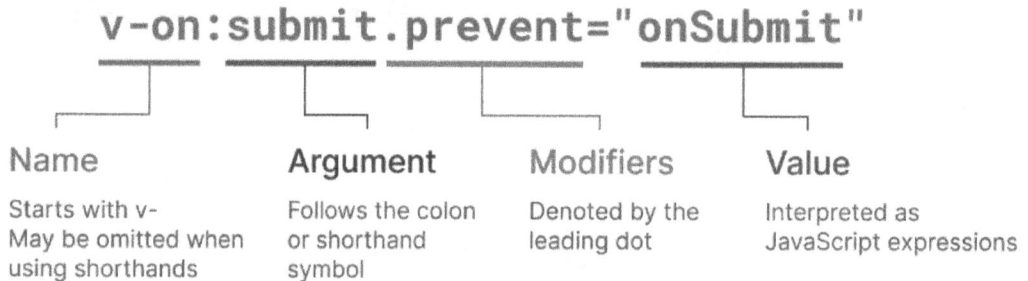

$$\texttt{v-on:submit.prevent="onSubmit"}$$

Name	Argument	Modifiers	Value
Starts with v– May be omitted when using shorthands	Follows the colon or shorthand symbol	Denoted by the leading dot	Interpreted as JavaScript expressions

Figure 6.3: Breakdown of syntax for a Vue directive

A directive in Vue.js is made up of four main sections:

- The name of the directive – for example, `v-on`, `v-if`, or `@` if using the available shorthand.
- An argument, such as `click` for a `click` event, or the name of a prop.
- A modifier such as `prevent` or `stop`.
- The actual value required by the directives. In the case of an event handler, the value would be the function that is run when the event is triggered, while in the case of props, the value would be a simple variable.

Event modifiers are appended to the event directive, so, for example, for `preventDefault`, you would write `@click.prevent="methodName"`.

Learning about all of these modifiers is out of the scope of this book as some of them (such as `capture` and `passive`) have very specific usage that is not always required. If you want to learn more, you can get more details in the Vue.js official documentation (`https://vuejs.org/guide/essentials/event-handling#event-modifiers`).

Before moving on, we are going to try and use one of these modifiers in our application: `.once`. This modifier prevents our event handler from being triggered more than once. This is very useful when you have actions that can just be performed once, for example, saving a new entry or updating a table row.

We are going to apply this to the same `click` event that we recently wrote. By doing so, we will just be able to update the current time once, as clicking the button multiple times will not perform any action:

```
<button @click.once="onUpdateTimeClick">Update Time</button>
```

The usage of the modifier is quite simple, as we have just shown. All we needed to do to prevent the event from being triggered more than once was add the word `.once` after our event directive.

> **Modifiers are not just for events but for all directives**
>
> The ability to provide a modifier to a directive, as shown previously, is not exclusive to events; in fact, modifiers are a feature available to directives in general. The only built-in directive to have modifiers is v-on, but you can create your own custom directive or use an external package that could have its own modifiers.

After this brief introduction of native events, it is time to learn about custom events and how they can be used to broadcast messages between components.

Connecting components with custom events

Frameworks such as Vue.js allow us to break down our application into very small components, sometimes as small as a single HTML element. This would not be possible without a strong communication system.

Native HTML elements offer events as a way for them to trigger actions and pass information to their parents, and Vue.js components use a very similar pattern by offering custom events.

Custom events are not something new, as they have been around JavaScript for quite some time, but they are usually very verbose to use in vanilla JavaScript, whereas with Vue.js, creating, emitting, and listening to custom events feels easy and intuitive. Let's see how these are defined and used.

To learn about custom events, we are going to modify the Companion App by adding the ability for the user to delete a post:

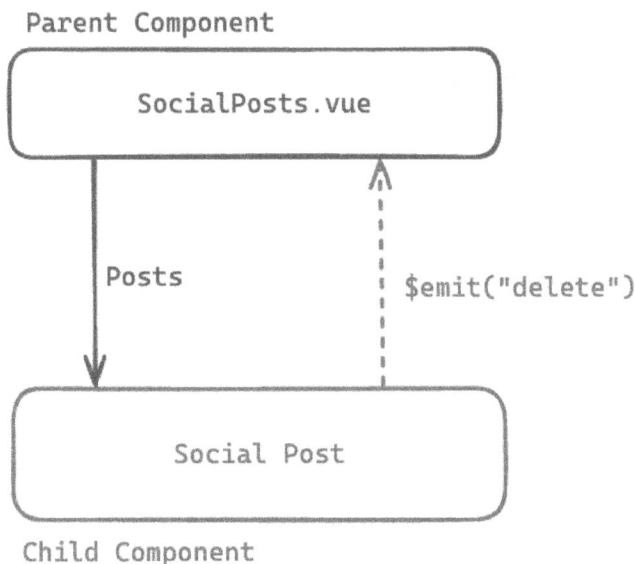

Figure 6.4: Diagram displaying the flow of data between parent and child components

As visualized by the preceding diagram, properties such as `Posts` are used by the parent to send information to the child. On the other hand, events are used for the child to emit information to the parent.

In the example illustrated in *Figure 6.4*, the `SocialPosts.vue` file sends a list of social posts to the child, and the child uses a custom event called `delete` to ask the parent to delete a post.

Let's see how to implement this custom event in the `SocialPost.vue` file. The first step requires us to trigger a native `click` event when the *delete* icon is clicked. Just as we did in the previous section, we achieve this by adding the `@click` directive in the HTML:

```html
<div class="header">
  <img class="avatar" :src="avatarSrc" />
  <div class="name">{{ username }}</div>
  <div class="userId">{{ userId }}</div>
  <IconDelete @click="onDeleteClick" />
</div>
```

We then add the `onDeleteClick` method in the `<script>` tag:

```js
...
onMounted( () => {
  console.log(props.username);
});
const onDeleteClick = () => {}
</script>
```

Next, we need to define the event and trigger it when `onDeleteClick` is triggered:

```js
...
onMounted( () => {
  console.log(props.username);
});
const emit = defineEmits(['delete']);
const onDeleteClick = () => {
  emit('delete');
}
</script>
```

We define emits using a compiler macro called `defineEmits`. This follows the same syntax of `defineProps` as it accepts an array of events that can be emitted by the component. In our case, we just defined one event of `delete`.

`defineEmits` returns a function that can be used to trigger our events. It is suggested to name this constant `emit` to align with the native `$emit` method available in Vue.js.

This function accepts two arguments. The first is the event name, and the second is just a parameter that will be passed to whoever listens to the event. In our case, this was defined with `emit('delete')`, where `delete` was our custom event name.

Now that the child custom event is fully set, it is time to ensure that the parent is listening to the event. Let's go over `SocialPosts.vue` and put our skills to use:

```
    . . .
    :retweets="post.retweets"
    :key="post.userId"
    @delete="onDelete"
  ></SocialPost>
</template>
<script setup>
import { reactive } from 'vue';
import SocialPost from '../molecules/SocialPost.vue'
const onDelete = () => {
  posts.splice(0, 1);
}
. . .
```

Listening to a custom event in Vue.js is no different from listening to a native event. In fact, both use the `v-on` (or `@`) directive, as shown by `@delete="onDelete"`. This code, just like other event listeners, will wait for the event to be triggered before running the provided function.

The method triggered by the event is called `onDelete` and uses plain JavaScript to remove the item from the `Posts` object. If we tried the current code, we would see that the functionality works as expected and the first post is deleted when the *delete* icon is clicked.

This is great but not perfect, as we would like to choose which post to delete instead of hardcoding this to be the first post. To fix this bug, we would have to add an argument to define which one is the correct post to delete when declaring our custom event.

Event arguments

Until this moment, all the events that have been handled in the Companion App required access to neither the event object nor the external argument.

Now, we are going to learn how you could expand your event handler to accept these extra values.

Since the methods attached to an event have access to all reactive data of a component, you may think that passing arguments may not be extremely useful, but this is not the case.

In fact, arguments are extremely helpful when triggering an event from an element that is using `v-for` – just like in our case, in which we need to inform our event of the `post` index. Having access to the correct value of the post that is triggering the event is vital for the functionality to run correctly.

To be able to delete the correct post, we first need to add the value of the `post` index when triggering the event handler and then ensure that this value is read and used by the method.

1. Let's make the `post` index available by exposing it within `v-for`:

```
<SocialPost
  v-for="(post, index) in posts"
  :username="post.username"
```

2. Next, add the index to the event declaration:

```
<SocialPost
  v-for="(post, index) in posts"
  :username="post.username"
  ...
  :key="post.userId"
  @delete="onDelete(index)"
></SocialPost>
```

3. Finally, add the `postIndex` argument in the event handler:

```
const onDelete = ( postIndex ) => {
  posts.splice(postIndex, 1);
}
```

After these three modifications, we should be able to delete the correct post from our Companion App.

Getting access to the events native object with $event

Event handlers have access to a special argument called `$event`. This is passed automatically to all events that have no argument or can be passed directly from the HTML when passing custom parameters: `<button @click="handler($event, customParameter) />"`.

Summary

In this chapter, we have revisited previously learned topics by analyzing the changes in the Companion App. We then expanded our knowledge of props by introducing all their possible options. Next, we moved on to event handling by explaining how to use native HTML events within Vue.js components before closing the chapter with the definition of a custom event handler and how this can be used to flow information between components.

You should now be able to create and use custom events. With this knowledge, plus the additional knowledge on props, you should be able to work on larger applications that have multiple components. You should also make use of the changes that have been made in the Companion App to learn how to navigate the new code base.

In the next chapter, we are going to continue to build on the topic of "increase scope" by removing the hardcoded values, starting to use external APIs, learning how to handle side effects by watching for value changes, and overcoming the limitation of computed properties. These changes will increase the complexity of our application even further, while at the same time, providing us with a vital skill to increase our knowledge of the Vue.js framework.

7

Handling API Data and Managing Async Components with Vue.js

In *Chapter 6*, we focused on how components can communicate with each other with the use of properties, which are used for parent-to-child communications and events to handle messages sent from a child to a parent.

In this chapter, we will stay on the topic of communication by showing how to communicate with an external source, such as an API. External communication is a very common method when developing a dynamic website that cannot make use of static data, and learning how to manage asynchronous operations will not only result in a clean user experience but also help keep the application performant.

Loading data from an external source such as an API makes data handling more complicated. In fact, when the data is hardcoded, we do not have to worry about anything, as the information is immediately available, while when working with data that comes from an external source, we need to not only think about the empty state that the app will be while the data is loaded but also consider the possibility of the data failing to load.

We will start the chapter by removing the hardcoded posts and loading them dynamically; we will then do the same with the comments by making the data on-demand. We will then enhance our application to automatically load more posts, using watch. Finally, we will learn how to define and use asynchronous components using <Suspense>.

The chapter will cover the following topics:

- Loading data from an API using the Vue.js life cycle
- Watching for changes in components using watch functions
- Handling asynchronous components with Suspense

By the end of the chapter, you will have learned how to load data and components dynamically. You will know how to create components that load data on demand and what benefits this brings to our application. You will also be able to handle side effects using watch and, finally, define and take care of asynchronous components to ensure your application is rendered correctly.

Technical requirements

In this chapter, the branch is called CH07. To pull this branch, run the following command or use your GUI of choice to support you in this operation:

```
Git switch CH07.
```

As part of this chapter, we are also going to use an external resource called Dummyapi.io. This website will provide a dummy API that we will use to load our post dynamically. To be able to use the API, you need to register and generate an **APP ID**. Creating an **APP ID** is completely free and can be obtained by creating an account on the following link (https://dummyapi.io/sign-in).

> **Note**
>
> Unfortunately, since the book was initially published, DummyApi is not being maintained anymore. It is currently not possible to register new accounts and therefore generate new APP IDs. To be able to continue your studies, you can utilize the following APP ID: 6835b3b89ccb3c0e9f8e75a1.

This new branch, CH07, includes just a couple of style changes and the replacement of the native button with the custom button component that we created in the last chapter, TheButton.vue.

The code files for the chapter can be found at https://github.com/PacktPublishing/Vue.js-3-for-Beginners.

Loading data from an API using the Vue.js life cycle

It is common for most of the applications that are built for the web to expose a level of dynamic content. Providing the ability to load information on the fly has been one of the most important factors that led to the growth of JavaScript frameworks, such as Vue.js.

Until now, the Companion App has been built using static data that is loaded directly within the components. Hardcoded values are not very common in real applications, and the posts and comments used within the application were just a stopgap to help us focus on the basic features of Vue.js, but it is now time to learn how to load data dynamically.

Being able to successfully handle asynchronous data load is very important. No matter how big or small your next application will be, it is very likely that you will be required to handle external data.

In this section, we are going to update two parts of our application. First, we are going to update `SocialPosts.vue` to load the post from an external source, and then we are going to change `SocialPostComments.vue` to also load comments dynamically but with a little twist, as we will implement something called "loading data on demand." We will then briefly discuss the implications that dynamic loading can have on performance and the user experience of our application.

Loading social posts from an API

Until now, the posts that our application displayed were always the same due to the hardcoded array of `posts` defined in `SocialPosts.vue`. In this section, we are going to use a public API offered by **DummyAPI** (`https://dummyapi.io/`) to make our post dynamic.

Services such as **DummyAPI** are very useful to develop application scaffolding and to help you practice your skills. There are plenty of free services like this one available on the internet, and they can easily be found using a search engine.

> **Research is part of development**
>
> Spend a couple of minutes navigating through the **DummyAPI** website and try to understand how we will use the API and what endpoints we will use. Learning external resources is a very important part of web development.

Loading external data will be achieved using Vue.js methods, the native Fetch API, and Vue.js life cycles. First, we are going to remove the old, hardcoded data from `SocialPosts.vue`. This file can be found in the `molecules` folder, as it is a component that renders a big section of our Companion App homepage:

```
const posts = reactive([]);
```

Then, in the same file, we are going to create a method that calls the external API to fetch our new posts:

```
const fetchPosts = () => {
  const baseUrl = "https://dummyapi.io/data/v1";
  fetch(`${baseUrl}/post?limit=5`, {
    "headers": {
      "app-id": "1234567890"
    }
  })
    .then( response => response.json())
    .then( result => {
      posts.push(...result.data);
    })
}
```

The preceding code uses the native JavaScript `fetch` method (https://developer.mozilla.org/en-US/docs/Web/API/Fetch_API) to send a `GET` request to the `dummyapi.io` API. Due to the API requirements, we need to pass `app-id` with the request. This can be obtained for free, as mentioned in the *Technical requirements* section.

We then fetch the result in the `json` format using `response.json()` and, finally, append the returned data to the post's `Reactive` property.

After the method has been defined, it is time to "call" it. When triggering an `async` request like in this case, we take advantage of the Vue.js life cycle to ensure that our request is triggered at the correct time.

In *Chapter 2*, we introduced the different life cycles and mentioned that the `created` life cycle is the correct one for asynchronous data. Our description of it was as follows:

> *"[The `created` life cycle] is the perfect stage to trigger asynchronous calls to gather some data. Triggering slow requests now will help us save some time, as this request will continue behind the scenes while our component is being rendered."*

Let's go and call our newly created method, `fetchPosts`, during the created life cycle. In contrast to other life cycles such as `mounted`, `created` does not need to be explicitly defined. The explanation from the official documentation is as follows:

> *"Because `setup` is run around the `beforeCreate` and `created` lifecycle hooks, you do not need to explicitly define them. In other words, any code that would be written inside those hooks should be written directly in the `setup` function."*

This simplifies our requirement, meaning that all we need to do is call the method after it is defined within the body of our component's JavaScript logic:

```
const posts = reactive([]);
const fetchPosts = () => {
    ...
}
fetchPosts();
```

At this stage, our posts should be dynamically loaded from the API, but the work is not done yet; in fact, the application displays the posts incorrectly:

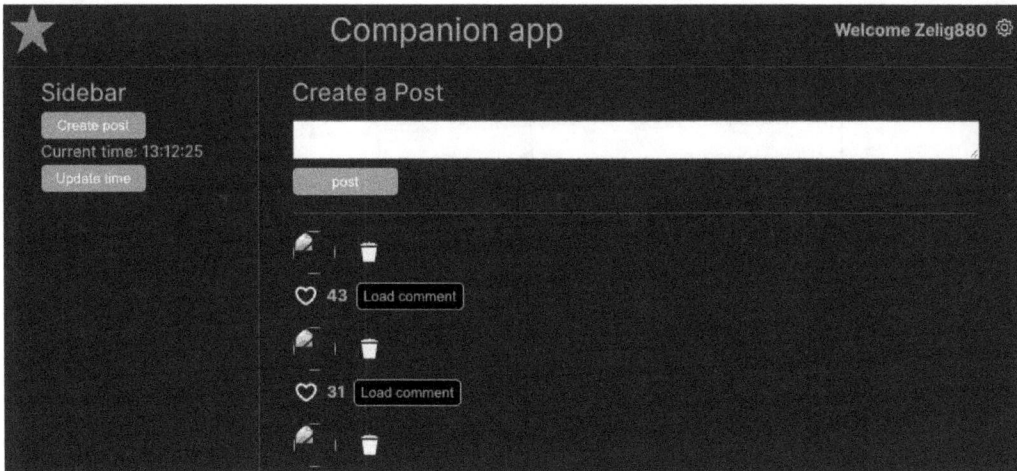

Figure 7.1: The Companion App displaying a broken UI

What caused the preceding error?

Before reading the answer, why don't you try to investigate what could have caused the issue with the rendering displayed in *Figure 7.1*? How would you go about fixing it?

The data fetched by the API is loaded and applied to our "posts" successfully, but the structure of the data does not match the one that we previously had set. This problem is related to the advanced properties settings that we just learned about in the previous chapter.

Fixing the SocialPost.vue properties' misalignment

Misaligned props are something that is very common in real applications, but they can be avoided. In fact, the reason why the app is rendering a broken UI is that we have not specified which properties are expected to be "required" in `SocialPost.vue`, and as a result, Vue.js tries to render the app with the data it has, resulting in the missing data being set as `null`.

Let's compare the previous hardcoded structure of the posts array with the new structure received by the API to see how the two structures compare and what changes are required to ensure the Companion App can correctly render the post information:

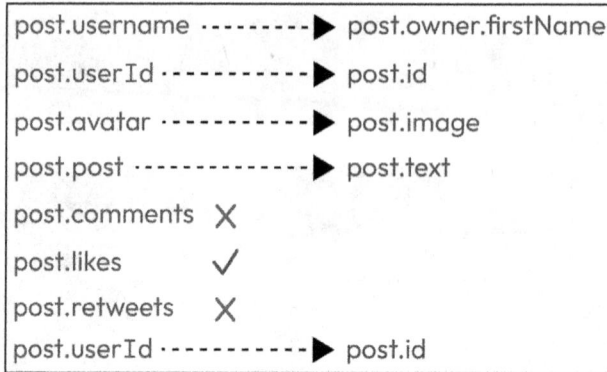

Figure 7.2: The transition between the previous structure of posts and the new one provided by the API

Figure 7.2 displays the changes we are going to make in our application to align with the new data. Some of the fields need to change to match the new object properties, and comments and retweets have been removed altogether. Calling the <SocialPost> component will now be changed to the following:

```
<SocialPost
  v-for="(post, index) in posts"
  :username="post.owner.firstName"
  :id="post.id"
  :avatarSrc="post.image"
  :post="post.text"
  :likes="post.likes"
  :key="post.id"
  @delete="onDelete(index)"
></SocialPost>
```

The preceding highlighted code shows the difference between the old and new component instances. Note that we had to replace userId with just id, which will be required later. It is now time to modify the child component to make sure it can function with the new data. This will involve a couple of steps:

1. Remove UserId from the UI, as it is too long:

   ```
   <div class="name">{{ username }}</div>
   <div class="userId">{{ userId }}</div>
   <IconDelete @click="onDeleteClick" />
   ```

2. Remove comments and retweets from the props declaration:

   ```
   const props = defineProps({
     username: String,
     userId: Number,
     avatarSrc: String,
     post: String,
   ```

```
    likes: Number,
    comments: Array,
    retweets: Number
});
```

3. Refactor `interactions` and `commentsNumber` from the UI and replace them with just `likes`:

```
<div class="interactions">
  <IconHeart />
  {{ interactions }}
  <IconCommunity />
  {{ commentsNumber }}
  {{ likes }}
```

4. Remove the logic associated with `interactions` and `commentsNumber`:

```
const commentsNumber = computed( () => {
  return props.comments.length;
});

const interactions = computed( ()=> {
  const comments = props.comments.length;
  console.log(comments, props.likes, props.retweets);
  return comments + props.likes + props.retweets;
});
```

Remove the condition attached to the Show comment button:

```
<TheButton
  v-show="hasComments"
  @click="onShowCommentClick"
  value="Show comment"
  width="auto"
  theme="dark"
/>
```

The preceding changes may seem quite complicated when reading them out, but they follow a logical pattern. They are all connected to each other, a modification in one component may result in a change in another and so on. For example, removing the `retweet` props in the parent would then result in the props being removed from the child component within `defineProps`, and consequently, the removal of any code logic attached to those props, and finally, any use of the props in the component template.

As you get more familiar with the framework, the preceding changes will feel trivial, but I have added them here and covered them step by step on purpose to give you some idea of how to logically think about a component and its data flow.

At this stage, the application should render correctly, and the home page should display posts coming from the dummy API we have implemented:

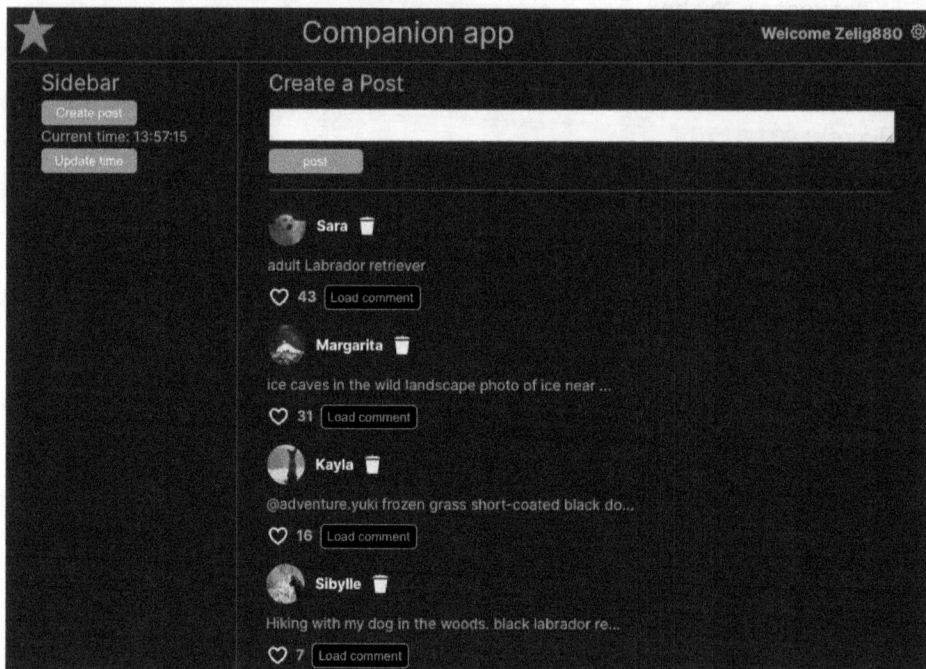

Figure 7.3: The Companion App displaying posts from the external API

The Companion App displays the post correctly again, but something is still missing. The logic that was used to display the comment does not work correctly anymore.

Before we move forward, you should improve the component that we just made to make it more reliable, by improving its properties.

> **Your turn – improve the properties set in SocialPost.vue**
>
> Spend a few minutes to enhance the props set in `SocialPost.vue`. Look at each props and decide if they should be required, by using the `required` attribute on props or if they could be set to optional by defining a `default` value.

Load comments on demand

When the `posts` data was hardcoded, the comments associated with the individual post were available on the first render, and we were able to pass them immediately between the parent and child components. But now that the information is loaded dynamically, we can change our logic to just load the comments on demand.

Imagine an application where all the data is loaded on the main page and has to be passed around between tens of components. No matter how hard you would work, the code would be quite hard to maintain. Passing properties through many layers of components is known in the industry as "props drilling."

Props drilling can be avoided using two techniques:

- Loading data on demand
- Using state management

In this chapter, we are going to cover the first technique, loading data on demand, while state management will be covered later in *Chapter 11.*

Making a change like the one we just performed prevents some data from being loaded immediately, with it just being fetched on demand. This is a very good practice to improve both performance and code structure.

When analyzing how comments behave in our application, we can see that it would be quite wasteful to load all comments associated with the posts immediately, when we know that the application will just display them on user interaction (a button click).

We are going to work on two files. First, we are going to replace the `comments` property from `SocialPost.vue` with the post ID. Then, we will create the functionality required to load the comments on demand in `SocialPostComments.vue`.

At this point, you should have enough knowledge to remove the `comments` property from the component and replace it with another property, called `post-id`. This will be used later to request the correct comments from the dummy API, and the code changes should be the following:

```
<SocialPostComments
  v-if="showComments"
  :comments="comments"
  :post-id="id"
  @delete="onDeleted"
/>
```

Because the `comments` property was already removed from `defineProps` in the previous section of this chapter, all that was left to do was to ensure that we removed the prop from the `<SocialPostComments>` component declaration and replaced it with one called `post-id`.

The next step requires us to rewrite the logic that handles the load of the comments. We will start by modifying the properties, by removing the `comments` one and replacing it with the newly passed `post-id`:

```
<script setup >
  import IconDeleteVue from '../icons/IconDelete.vue';
```

```
    const props = defineProps({
      comments: Array,
      postId: String
    })
</script>
```

The ID of the post is required for us to be able to fetch the correct comments from the API. It is very common when passing IDs around to prefix them with the actual context. So, in our case, instead of naming our prop ID, we named it `postId`. Small improvements such as this can really help keep your code clean and readable.

> **Defining multiword props using kebab case**
>
> Did you notice that the property we just defined is called `postId` in the component declaration but is passed in the component using kebab case, `post-id`? HTML is not case-sensitive, so using `postId` would just mean the same as `postid`. Therefore, to ensure that we better define the multiword props, we use kebab case – that is, "word-word."

It is now time to make the necessary changes to ensure that `SocialPostComments.vue` works correctly with the new posts structure and create the logic required to load the comments on demand.

We are going to use the following path from the API, `/post/{postId}/comment`. This is going to return us the comment from a given `postId`, where `postId` is the actual ID of the post selected.

Let's break down all the changes required:

1. First, we import `reactive` from `vue` and use it to define a new array for `comments`:

    ```
    import { reactive } from 'vue';
    const props = defineProps({
      postId: String
    });
    const comments = reactive([]);
    ```

2. Second, we create a new method called `fetchComments` that accepts a parameter of `postId`:

    ```
    const fetchComments = (postId) => {  }
    ```

3. Next, we create a fetch request in the body of the newly created method and use `postId` to create the correct request URL. Just like before, we make sure to pass the correct `app-id`:

    ```
    const fetchComments = (postId) => {
      const baseUrl = "https://dummyapi.io/data/v1";
      fetch(`${baseUrl}/post/${postId}/comment?limit=5`,
      {
        "headers": {
    ```

```
        "app-id": "1234567890"
      }
    });
  }
```

4. Then, we fetch the response JSON and assign it to the `comments` reactive:

```
fetch(...)
  .then( response => response.json())
  .then( result => {
    Object.assign(comments, result.data);
})
```

5. Finally, we call this function when the component loads. As mentioned before, calling a function within the script setup body is equivalent to calling it on the created life cycle. The function will receive the `postId` property as its argument:

```
const fetchComments = (postId) => {
  ...
};
fetchComments(props.postId);
```

At this stage, our Companion App will render the `comments` body that is currently received from the API request:

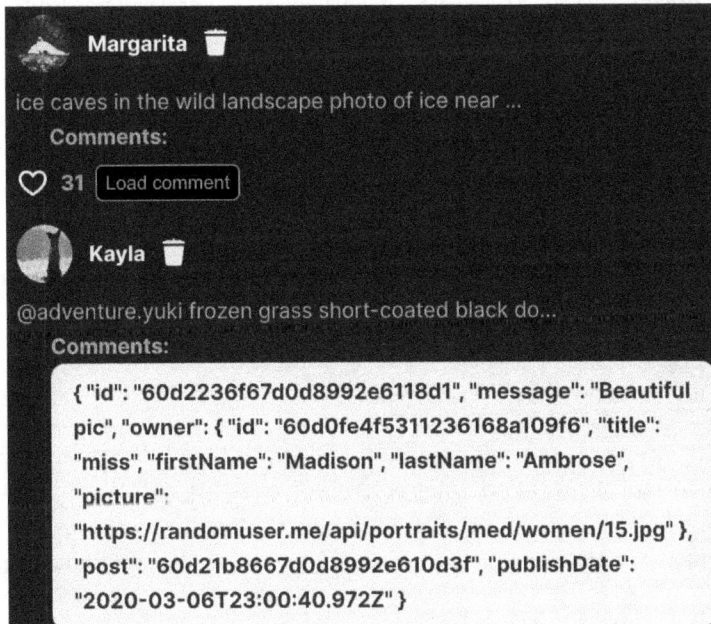

Figure 7.4: The comments body displayed in the Companion App

From the screenshot displayed in *Figure 7.4*, we can deduce that there are two main changes that we need to implement in the component:

- Improve the UI when no comments are available

- Format the body of the comments to just show the author's name and the message, instead of the raw object received by the API

To improve the user experience and show a different message if there is a post with no comments, we can use the built-in `v-if` and `v-else` directives:

```
<template v-if="comments.length === 0"></template>
<template v-else></template>
```

Just like a simple `if`/`else` statement, when a Vue.js component receives elements that include a `v-if` and `v-else` directives, it will just render one of the two, depending on the condition received. In our case, the first block will be rendered if the comments array is empty and has a length of zero, while the second block will render if comments are available.

> **Use <template> to avoid unused HTML elements**
>
> You may have noticed that we used an element called `<template>` when introducing the `v-if/v-else` code block. The `<template>` element is a special Vue.js element that allows you to add logic without the need to add an HTML element. In fact, if the template element was not available, we would have had to add `<span>` or `<div>` just to allow us to add our logic. Every time you use the `<template>` element, it will disappear and will not render anything within the DOM. It is very useful in logic that uses `v-if`, `v-else`, `v-if-else`, or `v-for`.

Let's fill in the code blocks that we just created with the correct HTML. The first block will just show a static message for the empty state, while for the second, we need to analyze the object received by the API and understand what we want to display.

It is now time to focus on the structure of the comments. The `comment` object seems to include multiple properties, but the only ones we should use are the name of the user and the message, therefore `owner.firstName` and `message`, respectively:

```
<template v-if="comments.length === 0">
  There are no comments for this post!
</template>
<template v-else>
  <p>Comments:</p>
  <div v-for="{owner, message} in comments" class="comment">
    <p>{{ owner.firstName }}: <strong>{{ message }}</strong></p>
  </div>
</template>
```

The `v-if` statement includes a simple static message, while `v-else` contains a loop created using the `v-for` directives, `<div v-for="{owner, message} in comments" class="comment">`, that includes two mustache templates to format our string, `<p>{{ owner.firstName }}:` `<strong>{{ message }}</strong></p>`.

After the latest modification, our Companion App should display a nicely formatted layout for our comments.

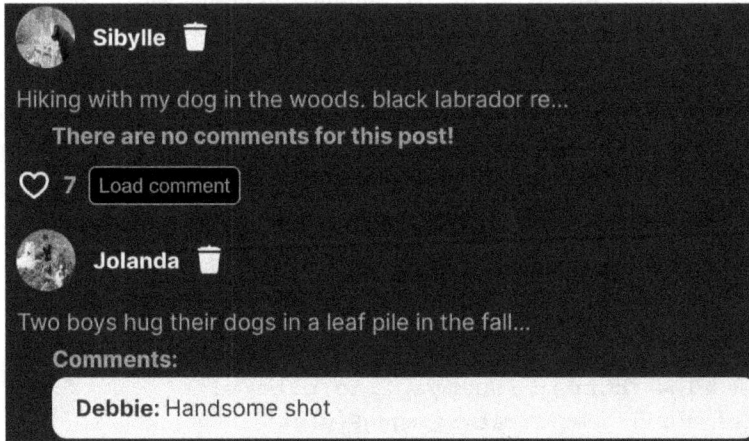

Figure 7.5: The Companion App's formatted comments

In this section, we learned how to load information asynchronously using an external API; we then walked through the changes necessary to ensure our application can work with the new dynamic data. The section also included a few tips to help you improve your Vue.js skills, such as the use of `<template>` to keep the HTML clean, the need to declare multi-word properties with kebab case, and the defining good properties to ensure that our components will not render incorrectly if any of the values are missing.

In the next section, we are going to learn how to trigger side effects, such as API requests, when watching for components' data changes.

Watching for changes in components using watch

In the previous section, we learned how to load our data dynamically by triggering an API request during the component rendering cycle. In this section, we will see another aspect of asynchronous data loading by describing how to handle API requests, triggered as a side effect when watching for data changes.

As we learned in *Chapter 5,* computed properties are a great asset to watch other properties and internal data to create new variables, but they are not handy when we need to trigger a side effect, such as a DOM change or an API call.

This is where a Vue.js feature called `watch` comes to the rescue. `watch`, just like `computed`, allows you to listen to any changes that occur to properties, reactive data, or another computed property, but it also provides you the ability to trigger a callback when data changes.

> **watch is just for edge cases, not everyday use**
>
> If you find yourself using `watch` often during your development, it means that you are using `methods` and `computed` incorrectly. It is very common for inexperienced Vue.js developers to overuse `watch`. I have personally used it just a handful of times in years of development.

We are going to change our application to automatically load more posts if there are fewer than four posts shown on screen. As mentioned previously, we are going to watch for a component variable (in our case, the `posts` array), and we are then going to trigger a side effect (in our case, another API call) when a certain condition is met.

Let's develop this together. First, we need to update our `fetch` method to accept a page parameter to ensure that we fetch new posts and not always the same ones:

```
const fetchPosts = (page) => {
  const baseUrl = "https://dummyapi.io/data/v1";
  fetch(`${baseUrl}/post?limit=5&page=${page}`, {
    "headers": {
      "app-id": "1234567890"
    }
  })
    .then( response => response.json())
    .then( result => {
      posts.push(...result.data);
    })
}
```

The `fetchPosts` method now accepts a parameter of `page` and appends it to the request query parameter, `` `${baseUrl}/post?limit=5&page=${page}` ``.

We are then going to create a new ref called `page` that will hold the current page value. To accomplish this, we are going to import `ref` from Vue.js:

```
import { reactive, ref} from 'vue';
```

Next, we define and initialize the variable. Because the first page of the posts is going to be 0, we are going to use this number as the initialization value:

```
const page = ref(0);
```

Lastly, we are going to pass this `ref` when calling `fetchPosts`:

```
fetchPosts(page.value);
```

Remember that because we used `ref` to define our page variable, we need to use the `.value` notation to access its value.

Now, it is time to create our `watch`. Just like `computed`, `watch` will depend on one or more other reactive values, like `Ref` and `Reactive`, and will include a callback value. The syntax is `watch( dependentData, callback(newValue, oldValue))`. So, in our case, the code would look like this:

```
<script setup>
import { reactive, ref, watch } from 'vue';
import SocialPost from '../molecules/SocialPost.vue'
watch(
  posts,
  (newValue, old) => {
    if( newValue.length < 4 ) {
      page.value++;
      fetchPosts(page.value);
    }
  }
)
```

`watch` is meant to observe variable changes and trigger a callback. In our example, the application observes the variable called `posts` and triggers a callback that will change the `page` variable and fetch new posts, using our API. The callback triggered by `watch` offers two values; the first is the new value that the observed variable has received, and the second is the old value.

Most of the time, you will probably just use the first value (therefore, the new value of the observed data), but having access to both values is very useful when the effect triggered is dependent on the "change" of the variable. For example, this may be needed if you have an animation that shows different effects, depending on whether the value increases or decreases. This would only be possible if you have both the old and the new values to compare.

Our Companion App should now be ready to be tested. To test our new functionality, delete a couple of posts by clicking the delete icon next to them, and then see the new posts load dynamically.

There are a few more options available with `watch`, such as the ability to trigger it after the effect has taken place in the DOM, or the possibility to trigger it immediately when the component renders. However, these are out of the scope of this book, as they are for more advanced use and would just be confusing at this stage.

Remember that using watch has a cost. In fact, there is a cost to be paid for an application observing a value, but there is an even greater performance cost in triggering a callback on every change of a value. For this reason, it is recommended to use watch only when needed and to ensure that the body of the callback is not too resource-intensive.

> **Use computed to minimize watch callbacks**
>
> If you are observing a variable that changes too often and want to try and improve on performance, you can use computed to create a new variable. This new computer property can then be watched instead. Because computed properties are cached, this approach is more performant.

That completes our introduction to watch. This is a useful feature that, when used correctly, can help you create clean and readable components.

In this section, we learned what watch is and how using it improves our component, giving us a chance to trigger side effects from our application, such as an API request or DOM modification. We then made changes in the Companion App to better understand this topic. Finally, we explained the drawbacks of using watch and discussed when and how to use it.

In the next section, we are going to introduce a built-in component called <Suspense>. This feature simplifies how we handle the loading state of asynchronous components.

Handling asynchronous components with <Suspense>

Handling dynamic data loading is never easy. In fact, when data is static and hardcoded, displaying the information takes no effort, as the data is available on first render, but when the data needs to come from an outside source, such as a database or a third-party API, then the complexity increases. When loading information asynchronously, the data is not available on the first load, forcing us to handle a "loading state" until the data is available, or having to display an error state if the loading event has failed to complete.

To prevent having to handle state changes in multiple components and code duplication, Vue.js introduced a globally available built-in component called <Suspense>.

With <Suspense>, we can orchestrate all loading states at once with very clean syntax. In the following section, we are going to first understand what makes a component asynchronous and then understand how this can be used to simplify our code.

> **An experimental feature**
>
> At the time of writing, <Suspense> is still an experimental feature, and there is no set date on when and if it will become a core part of the Vue.js framework. The team is fixing a few bugs with it and more importantly, developing server-side support before it is turned into a full feature.

Understanding asynchronous components

In the preceding introduction, we discussed an async component by using the example of a component that needs to load data dynamically, but that is not the actual definition of an async component.

An async component is "*a component that requires an asynchronous function to be performed and completed before the rendering can be initialized.*"

The important takeaway from the preceding definition is the words **asynchronous** and **completed**. The presence of both an `async` operation and the need for it to be completed are requirements for a component to be defined as an asynchronous component.

In fact, if we look at our current application, we can see that we already have dynamic data that is loaded in `SocialPosts.vue`, but this does not make it an async component, as the component is rendered immediately and does not wait for the `fetch` operation to be completed.

An asynchronous component is characterized by the presence of a top-level **await** function in the `<script setup>` code block.

> **An asynchronous component affects the page rendering**
>
> An asynchronous component will stop the rendering itself and all of its children until the data is fully loaded. This should just be used if the component has no reason to render without the data being available.

Let's look at our Companion App and try to find a good candidate to turn into an async component. Currently, only two components have asynchronous operations between them, `SocialPosts.vue` and `SocialPostComments.vue`.

If we look at the logic within the `SocialPostComments.vue`, we can see that the component does not currently function properly. The current component logic displays a message when the comments array is empty, `"<p>There are no comments for this post!</p>"`, but this message is also displayed when the component first renders. This is because the component renders immediately, even if the `fetch` request is still active.

This is a very good candidate to be turned into an async component. In fact, this component has both an "async operation" and the "need for it to be completed," as mentioned in the definition of an async component.

Turning a component into an asynchronous component

Vue.js provides a very simple way to define a component as an asynchronous component. In fact, all we need to do is ensure that the component includes one or more `await` functions within its body.

When the Vue.js framework sees an await function, it automatically defines the component as async.

Let's modify `SocialPostComments.vue` to await the `fetch` method:

```
<script setup >
import { reactive } from 'vue';
const props = defineProps({
  postId: String
});

const comments = reactive([]);
const fetchComments = (postId) => {
  const baseUrl = "https://dummyapi.io/data/v1";
  return fetch(`${baseUrl}/post/${postId}/comment?limit=5`,
  {
    "headers": {
      "app-id": "1234567890"
    }
  })
    .then( response => response.json())
    .then( result => {
      Object.assign(comments, result.data);
    })
};
await fetchComments(props.postId);
</script>
```

Our component logic required just two small changes to turn the component into a dynamic one. First, we made sure that our `fetchComments` method returned a promise by adding `return` before the `fetch` method. Then, we added `await` when calling the method. These two changes were all that was needed to ensure the component would turn into an async component.

All that is left to do now is learn how to use asynchronous components. In fact, at this stage, the Companion App is unable to load comments, and clicking on the "Load comment" button would log the following error:

```
▲ ▼ [Vue warn]: Component <Anonymous>: setup function returned a        SocialPost.vue?t=1702760833120:26
promise, but no <Suspense> boundary was found in the parent component tree. A component with async
setup() must be nested in a <Suspense> in order to be rendered.
```

Figure 7.6: An error message triggered by Vue.js when trying to incorrectly load an async component

The error message mentions the need for the async component to be nested in `<Suspense>` in order to be rendered. Let's learn what this `<Supense>` is and how it can be used to load async components.

Rendering async components

In the previous section, we made the `SocialPostComments.vue` file asynchronous, and it is now time to learn how to handle this component to ensure that it is loaded correctly.

As we previously mentioned, when a component is turned into an asynchronous one, it then requires us to handle its loading state. Loading this component normally, as we are currently doing in the application, would not work, as the component is not immediately available, so we need to find a way to handle its load gracefully.

As always, the Vue.js core team has worked hard to provide us with all the tools we need to quickly accomplish complex operations.

The component is defined on the official Vue.js documentation as follows:

> "<Suspense> *is a built-in component for orchestrating async dependencies in a component tree. It can render a loading state while waiting for multiple nested async dependencies down the component tree to be resolved.*"

Wrapping one or multiple components within <Suspense> prevents them from rendering until all the async operations are completed. Furthermore, <Suspense> also allows you to display a "loading" component while the async operations are completed.

<Suspense> is already preloaded in the application, and it does not need to be imported. Let's open SocialPost.vue and change our code to correctly load our async component:

```
...
<div class="post" v-text="post"></div>
<Suspense v-if="showComments" >
  <SocialPostComments
    v-if="showComments"
    :post-id="id"
    @delete="onDeleted"
  />
</Suspense>
<div class="interactions">
...
```

The use of this built-in component is very simple. In fact, all we need to do is wrap SocialPostComments within <Suspense>, as highlighted in the previous code block, and move the v-if directive, v-if="showComments", from SocialPostComments to the built-in Suspense component. After these changes, SocialPostComments will simply render after the async operation within the component resolves.

There are cases in which you may be required to display a loading indicator while the async operation completes. <Suspense> provides a named slot called **fallback** that can handle this. Let's learn how to use this feature by adding a fallback message while the post comments load.

To implement a fallback message, our code would require the following modifications:

```
<Suspense v-if="showComments" >
  <SocialPostComments
    :post-id="id"
    @delete="onDeleted"
  />
  <template #fallback>
    fetching comments...
  </template>
</Suspense>
```

To add a message while the comments are fetched, we use the **fallback** slot alongside our `SocialPostComments`. This slot is going to be defined using the `<template #fallback>` syntax. The content of this slot will just be displayed while the async operation within `SocialPostComments` is run, and it will disappear as soon as the component is rendered.

> **Custom error handling**
>
> At this stage, error handling is not handled by `<Suspense>`, and this needs to be handled manually using `onErrorCapture()` hooks. Explaining this is beyond the scope of this book.

Summary

We have now completed everything we had to learn about async data and component loading. We started the chapter by removing the hardcoded `posts` and replacing them with dummy data that is loaded dynamically. We then fixed the property mismatch caused by the data change, learning how to prevent this in the future by improving the use of property typing and validations.

Then, we learned how to change our data flow to load comments on demand and defined when this should be used, as well as the performance and user experience benefits that this brings. We then introduced another feature related to async operations, `watch`. We used this feature to trigger a side effect and automatically load more posts when the number of posts reaches a certain number.

Lastly, we learned how to create and handle asynchronous components. We described what makes an async component and changed our Companion App to ensure our comments were fetched before loading the component. We concluded the chapter by introducing the `<Suspense>` built-in component, using it to correctly load our asynchronous component, and we also examined a fallback feature that shows text while the component loads.

At this stage, you should be able to fully handle asynchronous data loading, side effects, and components that require JavaScript promises to be fulfilled before being displayed.

In the next chapter, we are going to move our focus away from Vue.js and focus on testing our application. We will learn the basics of end-to-end testing with **Cypress** and unit testing with **Vitest**.

Part 3: Expanding Your Knowledge with Vue.js and Its Core Libraries

At this stage in our journey, it is time to introduce the external libraries that are part of the Vue.js ecosystem that are required to build production-ready applications.

This part contains the following chapters:

- *Chapter 8, Testing Your App with Vitest and Cypress*
- *Chapter 9, Introduction to Advanced Vue.js Techniques – Slots, Lifecycle, and Template Refs*
- *Chapter 10, Handling Routing with Vue Router*
- *Chapter 11, Managing Your Application's State with Pinia*
- *Chapter 12, Achieving Client-Side Validation with VeeValidate*

8

Testing Your App with Vitest and Cypress

No matter how much experience you have under your belt in writing code, testing your app is a must to produce a high-quality code base. There are different testing tools out there, but in this book, we are going to learn Vitest for unit tests and Cypress for end-to-end testing.

Testing is a large topic, and in this chapter, we will learn the basics of both testing methodologies, leaving out advanced techniques for future reading and training that you will encounter in the course of your career.

First, we are going to learn the different testing methods and how they each contribute to producing quality software. We will then introduce unit tests by learning how to use Vitest and Vue Test Utils to test individual components in our application. Then, we will move our focus away from the individual components and concentrate on our application, by introducing **E2E (end-to-end)** tests with Cypress. We will then write an E2E test to cover a small user journey before moving on to the last section, which will be used to introduce advanced topics such as mocking and spies.

This chapter consists of the following sections:

- Testing pyramids
- Unit testing with Vitest
- E2E tests with Cypress
- Introducing advanced testing techniques

By the end of this chapter, you will have gained a basic understanding of testing in general. You will be able to set up both Vitest and Cypress on your future projects and know how to write basic tests in both unit tests and E2E tests. Finally, you will also be exposed to future testing techniques that you may encounter in your career.

Technical requirements

In this chapter, the branch is called CH08. To pull this branch, run the following command or use your GUI of choice to support you in this operation:

```
git switch CH08.
```

The branch includes all the files updated from the previous chapter.

The code files for the chapter can be found at `https://github.com/PacktPublishing/Vue.js-3-for-Beginners`.

The testing pyramids

Clean code, coding standards, and peer reviews are essential parts of a good application, but they are not the only ones. In fact, a good and reliable application is not only the source of good development but also the result of a good level of testing coverage available within the application.

Testing covers a very large spectrum. Some companies do very minimal tests, letting their end users be the actual testers by shipping them new code to look for bugs and errors, while others invest time and budgets in developing comprehensive sets of tests and adding them to their processes.

Even if companies invest different amounts of time in testing, all developers can at least agree that increased testing can, on average, result in fewer bugs being shipped to the user and a more flexible application.

Multiple levels of testing can be developed for an application, and they are divided into layers that together form a pyramid (hence the name *testing pyramid*).

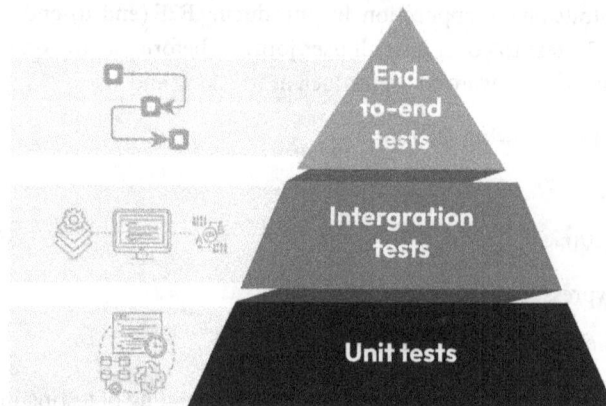

Figure 8.1: The testing pyramid

The bottom of the pyramid is where **unit tests** are found. Unit tests look at a single unit, which is usually a component or a helper file. These tests are very fast to run and, therefore, are usually produced in large quantities. Then, we have **integration tests**, which are where different parts of the system are connected to ensure that all parts work together. An integration test could test that the method successfully adds an entry to a database, by integrating the code and the database together, hence the name *integration test*. These tests take longer to run and require a bigger architecture, so they are expected to be used less than unit tests. Moving up in the pyramid, we find E2E tests. E2E testing encapsulates both manual tests, that are run by a quality assurance team, and automatic E2E tests that makes use of tools such as Cypress or Playwright. which are run manually by a quality assurance team, and automatic E2E tests are run by a tool such as Cypress or Playwright. An individual E2E test spans a full user journey and, due to its large scope, you will just require a few tests needed to cover your full application.

For the scope of this book, we are going to cover unit tests and E2E tests. Integration tests will not be covered because their usefulness is gradually decreasing, due to the advancement of E2E tools.

The difference in the test types is not only defined by the different bugs that they can catch but also by their "development cost" and "speed to run." Let's see how different tests perform in the following diagram:

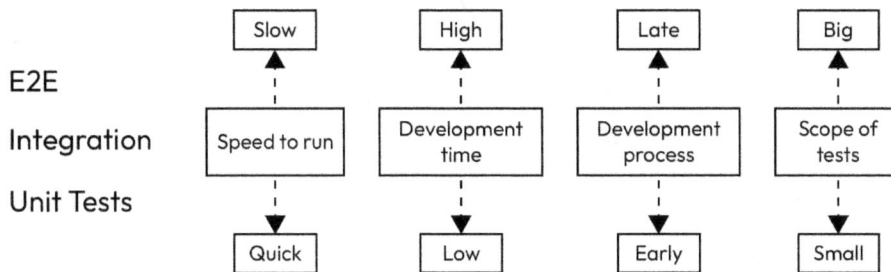

Figure 8.2: A comparison between the different test types

If an application only did E2E tests, it would catch bugs too late in development, slowing down the development process. Conversely, if an application only did unit tests, it would not pick up bugs and inconsistencies produced by different parts of the system working together, due to the fact that Unit Tests focus on a single unit and therefore cannot ensure that different part of the application work as expected.

Tests also exhibit differences in costs. A test cost is measured in the amount of time and effort that it takes to write the test, as well as its "speed to run," which refers to the amount of time a single test takes.

In an ideal world, all aspects of tests would be performed as part of your development process. But this is far from the truth. I have worked on many projects in my career, and I can count just a handful of businesses that really took ownership of their tests, implementing the whole pyramid. Unfortunately, for many businesses, tests are seen as an overhead (i.e, a cost that does not directly produce revenue).

Before moving forward to the next chapter and starting to develop our first test, it is important to try and define what makes tests worthwhile. We have already discussed the ability of tests to catch bugs early before they reach your clients. But this is not the only benefit; in fact, when asked by upper management why they should invest in tests, I usually prove my point by introducing the topic of "Increased flexibility and the reduced risk of change."

Increased flexibility and the reduced risk of change

When working on a big migration, the first task is ensuring that all E2E tests are set. Likewise, when developing a redesign, unit tests on the components are a must-have. A well-tested application does not only experience a lower level of bugs reaching the client but is also more flexible with updates. Knowing that you can easily test the major features of your application and quickly be informed if something has regressed or broken is a key aspect of agile development, setting your application for success and making it less risky to make changes to the code base.

I hope this is enough to convince you to understand the importance of testing. It is now time to move forward and learn how to write tests within our application. Tests can be quite tricky to implement on existing applications, and they also have a higher initial learning curve before you get started, but the payback of this investment is invaluable for your career and your code quality.

Let's see how Vitest can be set up and implemented to test some components of our Companion App.

Unit testing with Vitest

Vitest is one of the main unit test frameworks in the JavaScript ecosystem, and it has a syntax very similar to Jest. Most of the knowledge you will gain in this chapter is transferrable to other testing frameworks, as they all follow a very similar structure.

Vitest is already set up in our application, but we will cover the basic steps required to add it to your new or existing applications.

Installing Vitest in your application

The first step required is the package for **Vitest** and **Vue Test Utils**, which are, respectively, the test runner and the Vue components testing framework. We can achieve this by opening the terminal and running the following command in the root folder of our application:

```
npm install -D vitest @vue/test-utils
```

Now that the packages are installed, we just need to add a script to the package.json file so that we can simply run it in the future:

```
...
"scripts": {
  "dev": "vite",
  "build": "vite build",
  "preview": "vite preview",
  "test:unit": "vitest"
}...
```

The command string used is arbitrary, but it is common practice to name it `"test:unit"`. This will allow us later to also define `"test:e2e"` and clearly define the two different test procedures.

To run Vitest, we can just run the newly created script in our terminal with npm `run test:unit`. The result should be the following:

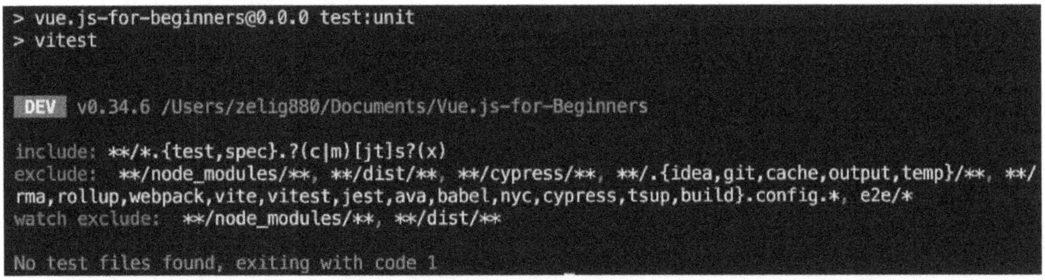

```
> vue.js-for-beginners@0.0.0 test:unit
> vitest

 DEV  v0.34.6 /Users/zelig880/Documents/Vue.js-for-Beginners

include: **/*.{test,spec}.?(c|m)[jt]s?(x)
exclude:  **/node_modules/**, **/dist/**, **/cypress/**, **/.{idea,git,cache,output,temp}/**, **/
rma,rollup,webpack,vite,vitest,jest,ava,babel,nyc,cypress,tsup,build}.config.*, e2e/*
watch exclude:  **/node_modules/**, **/dist/**

No test files found, exiting with code 1
```

Figure 8.3: A terminal screenshot of the test results

The console terminal should show an error message, **No test files found**. This is what we expected, as adding the package to the repository does not also add any tests, which need to be added manually. It is time to learn how to structure a test, and we will do so by testing our atoms component named `theButton.vue`.

Writing our first unit test

Now that everything is set, it is time to start and write our first test using our base button component.

A unit test has a very important goal – test a single unit of your application. In a framework such as Vue.js, a unit is defined as a component, a composable, or a store file.

When testing a unit of logic, you should focus on its functionality and not its static content, such as text.

Creating a text-focused test that just compare text values, such as headings and buttons, will not only provide any benefits to our application, but will also be hard to maintain , due to its flakiness (produced by the fact that the tests will have to change every time we change the copy of our component).

Before creating our test file, let's open `TheButton.vue` and see what part of this component could be tested:

```
<template>
<button
  :class="theme"
>
  {{ value }}
</button>
</template>
<script setup>
```

```
defineProps({
  value: {
    type: [String, Number],
    required: true
  },
  width: {
    type: String,
    default: "100px"
  },
  theme: {
    type: String,
    default: "light",
    validator: (value) => ["light", "dark"].includes(value)
  }
})
</script>
<style scoped>
...
</style>
```

Our component is a standard button and, as such, includes all the basic functionality that we would expect from one. A few test scenarios for this component could be the following:

- Checking that the component renders successfully, ensuring that there are no mistakes in the component structure

- Checking that the default styles load by default

- Checking that the theme props work to change the theme

These three initial tests are great for us to get started on this topic. As I said before, tests can have quite a steep learning curve, so it is good to start small and slowly build on that knowledge.

It is now time to create our test files. Test files are named in the nameOfFileTested.spec.js format, so in our case, it would be TheButton.spec.js.

We can create a test file in any folder we want, but it is good practice to decide a standard for the project and consistently follow it. In our case, we are going to create this file in the __tests__ folder. The full path of the file should look like this:

```
/src/components/__tests__/TheButton.spec.js
```

Before we start discussing the test structure, we need to import a few modules that will allow us to create the test. We are going to import the `expect`, `describe`, and `it` methods from Vitest, the `mount` method from `@vue/test-utils`, and finally, the component itself. The `import` statement at the top of our test should look like this:

```
import { expect, describe, it } from 'vitest'
import {mount} from "@vue/test-utils";
import component from '../atoms/TheButton.vue'
```

Now that our imports are set, it is time to start to learn how to structure a test. The structure of tests follows a callback approach, where each method has a callback to another method, and so on.

Even if this approach, also referred to as callback hell, is usually a sign of bad practice in JS, it is the best way to provide structure for your tests.

To write a well-structured unit test, we can use the Given, When, and Then syntax that answers the following three questions:

- What are we testing (Given)
- What is the scenario (When)?
- What is the expected outcome (Then)?

We can then use `Given`, `When`, and `Then` to create our own test cases by creating a simple sentence. So, in the case of our button, the sentence could be as follows: "Given `TheButton.vue`, when *it is mounted*, then it *should render properly*."

With the preceding sentence. we can now go and write our first unit test.

> **Write your test in words first**
>
> It is beneficial to break down your tests by writing them in sentences first. Being able to write down all the tests by just defining a single sentence for each test will help you define better tests and ensure that you cover your components fully. Reading and comparing a couple of sentences is simpler than doing the same after the tests are written.

To structure our tests, we will make use of the `describe` and `it` methods provided by Vitest:

```
describe('TheButton.vue', () => {
  describe('when mounted', () => {
    it('renders properly', () => {
      // Test goes here
    });
  });
});
```

A test uses the describe method to define the aforementioned Given and When syntax, and the it method to define the actual test (Then). A test can have multiple nested descriptions. Test names are very important because they are used by a testing framework in error messages, triggered when a test fails. Having a very well-structured name will help you save time if something breaks and your test fails. Let's try to run our empty test for the first time and see what the output is. To run our tests, we need to open the terminal, access the project folder, and run the following **npm** command:

```
npm run test:unit
```

The output of the preceding command should be the following:

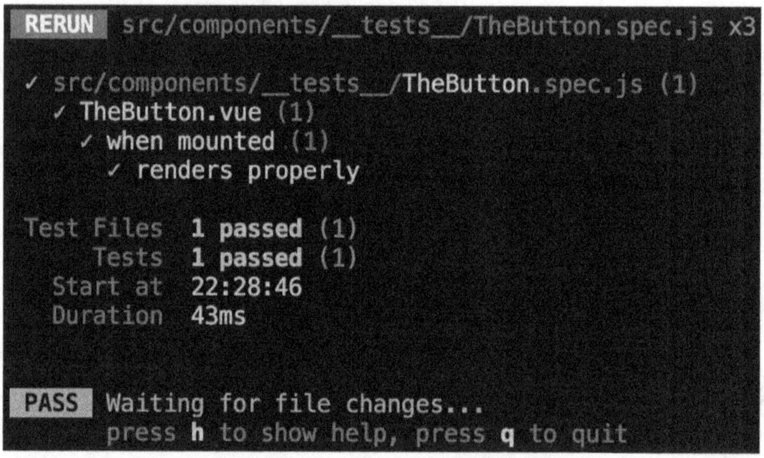

Figure 8.4: The terminal output of Vitest

The test results follow the same nested structure that we defined in our test. The words used within the describe and it methods form a readable word. If an error is triggered locally or in any deployment environment, the log will display the file name, followed by the different words we used to declare our test. In our case it would be *TheButton.vue when mounted renders properly.*

> **Vitest is fast – very fast**
>
> In the previous section, we mentioned that unit tests are fast, but we have not mentioned that among all unit test frameworks, Vitest is the fastest of all unit test frameworks. Its speed is to be attributed to the Vite server on which Vitest is built.

Our first test has not tested anything yet, as it has an empty body. Let's go back and see how we can test our component. To complete this task, we will use the mount method offered by Vue Test Utils and the expect method, used to tell the Unit Test engine what we want to test.

The mount method is used to initialize a component by rendering it within the test framework, while expect is used to define our test cases.

In every test, we start by setting out our component and any state that this may require. In our case, we just need to mount it:

```
const wrapper = mount(component, {});
```

Next, we will assert that our component rendered successfully by checking whether it includes a button element:

```
expect(wrapper.html()).toContain('button');
```

The expect method accepts the value being tested as its argument and is then chained to the test performed on the given value. Vitest comes with a huge list of assertions.

For example, if we wanted to create a test that checks if two number are equals, we would do the following:

```
const numberFive = 5;
expect(numberFive).toEqual(5);
```

The full test file should look like this:

```
import { expect, describe, it } from 'vitest'
import component from '../atoms/TheButton.vue'
import {mount} from "@vue/test-utils";

describe('TheButton.vue', () => {
  describe('when mounted', () => {
    it('renders properly', () => {
      const wrapper = mount(component, {});
      expect(wrapper.html()).toContain('button');
    });
  });
});
```

So, to recap, we learned that tests follow a very structured approach. They require us to define what we are testing, and the scenario and assertion that we are considering.

Let's add another test that would create the following test sentence: "Given TheButton.vue, when *it is mounted*, then it *defaults to the light theme.*"

As you can see from the first two parts of the sentence, Given and When are the same, so this means that we can reuse the existing code blocks and just add another test.

To check whether the correct theme is applied, we will check whether the CSS class light has been applied to the component.

```
import { expect, describe, it } from 'vitest'
import component from '../atoms/TheButton.vue'
```

```
import {mount} from "@vue/test-utils";
describe('TheButton.vue', () => {
  describe('when mounted', () => {
    it('renders properly', () => {
      const wrapper = mount(component, {});
      expect(wrapper.html()).toContain('button');
    });
    it('defaults to light theme', () => {
      const wrapper = mount(component, {});
      expect(wrapper.classes()).toContain('light');
    });
  });
});
```

As shown by the highlighted text in the preceding code block, adding extra tests that follow the same setup is quite simple. To complete the new test, we used `mount` again to initialize a version of our component. We then used `wrapper.classes()` to extract all the classes of the component and asserted this to contain `light` using `expect().toContain()`.

Due to the fact that both of the tests have the same When, we were able to reuse the describe method `describe('when mounted'..)`. Doing so will help us break up tests logically.

The tests are automatically updated, so when the file is saved, our terminal should output the new result with two passing tests:

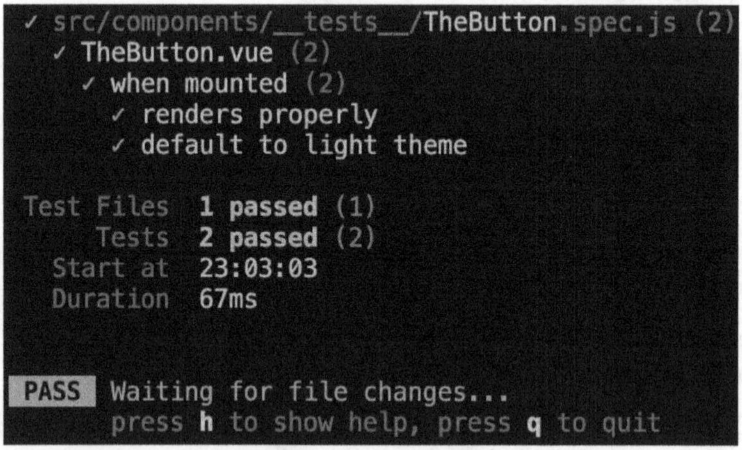

```
✓ src/components/__tests__/TheButton.spec.js (2)
  ✓ TheButton.vue (2)
    ✓ when mounted (2)
      ✓ renders properly
      ✓ default to light theme

Test Files  1 passed (1)
     Tests  2 passed (2)
  Start at  23:03:03
  Duration  67ms


PASS  Waiting for file changes...
      press h to show help, press q to quit
```

Figure 8.5: The Vitest test results showing two passing tests

> **Your turn**
>
> Write the last unit test to complete the testing of `TheButton.vue`. The last test should test that the component renders the dark theme when the correct properties are passed. Research Vue Test Utils and the Vitest documentation to learn how to define properties when mounting a component. Next chapter branch will include the test for you, so that you can check if your test was written correctly.

Unit tests are a big topic, and this was just a very simple introduction to give you some necessary initial information. Each project will create tests that are slightly different, either because of the project structure, the breakdown of components, or even the naming convention used. For this reason, the best way to truly learn how to write tests is to practice.

Before we move forward, I want to share one more tip to ensure you test your code correctly. In fact, it is common when writing unit tests to make them too dependent on the inside of a component. A unit test is expected to test what a single unit of code would **output** when given a specific **input**. What this means is that you should be able to test a component without the need to open the file, just by knowing the properties it accepts and the UI/events that it emits. When writing your unit tests, always focus on the input and output, and leave the implementation details out of your test.

In this section, we introduced unit tests and discussed how to use Vitest to test our component. We learned the syntax required to write a unit test and introduced Vue Test Utils to help us work with our Vue component. Then, we covered the importance of the test name and how this follows the Given, When, and Then methodology.

In the next section, we will ascend the pyramid and learn to write E2E tests. We will be able to reuse some of the knowledge we learned in this section and continue to learn how to make our code more reliable.

E2E tests with Cypress

It is time to move our testing focus from a micro level, offered by unit tests, to a macro level, provided by E2E tests. If unit tests focus on a single state of a single component, E2E will help us test a complete user journey.

As I mentioned previously, we will be able to use some of the knowledge that we learned in the previous section, since E2E tests follow a similar structure to the ones we learned about in Vitest.

Let's break down this section into three different parts. First, we are going to learn how to install **Cypress** in your application. Then, we will learn the basic structure of an E2E test file and its placement within a project. Lastly, we will end the section by writing an E2E test for our application.

The most used E2E tools within the JS industry are Cypress and Playwright. Cypress has been around for quite a few years, and it currently holds a very large section of the market. Playwright is a little newer, but it is gaining a lot of market share due to its integration with IDEs and extensive browser emulation.

Just like unit tests, the syntax of the different players on the market is very similar, so most of the features and syntax that you learn in one could be used in another one if you later decide to switch.

Let's start by installing Cypress into your project.

Installing Cypress into your project

Today's JS ecosystem offers us tools with amazing user experiences that offer very simple installation processes, and Cypress is one of those.

Our Companion App already includes a set of tests, and they are installed by the Vue project initialization that we introduced in *Chapter 3*. However, it is helpful to learn how to add Cypress to your new or existing project. Due to the fact that Cypress is already part of the Companion App, you would have to follow the following steps in a new folder.

The official documentation provided at cypress.io offers two installation methods, a **direct download** and an **npm installation**.

We are going to follow the installation using the npm package manager. To initialize this, access the root folder of your project in your preferred terminal and type the following command:

```
npm install cypress
```

The next step is to run Cypress. This can be achieved with the following npx command (note that the command is npx and not npm):

```
npx cypress open
```

After a few seconds, you should be presented with a welcome screen.

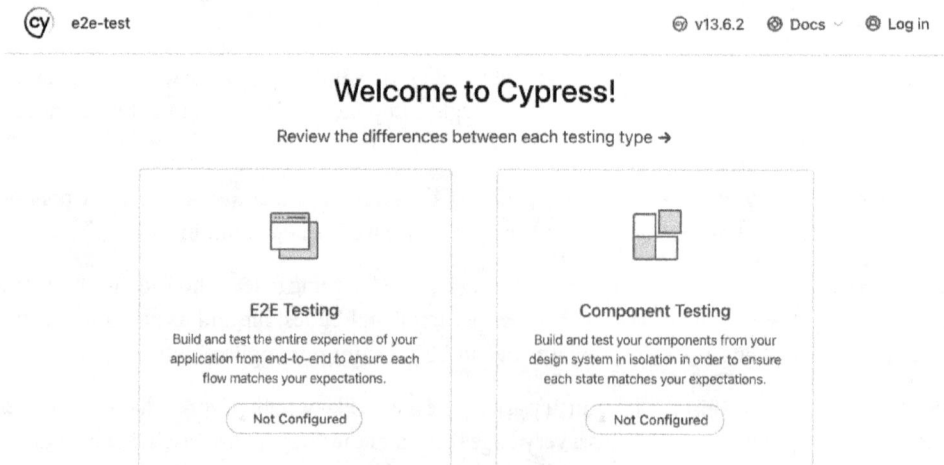

Figure 8.6: Cypress's opening dashboard

Cypress can be set up for E2E and component testing. The wizard in *Figure 8.6* marks our test as **Not Configured**. Let's begin the configuration of the E2E testing by clicking on the left block. Doing so will generate a set of example files that we will need to run our first test.

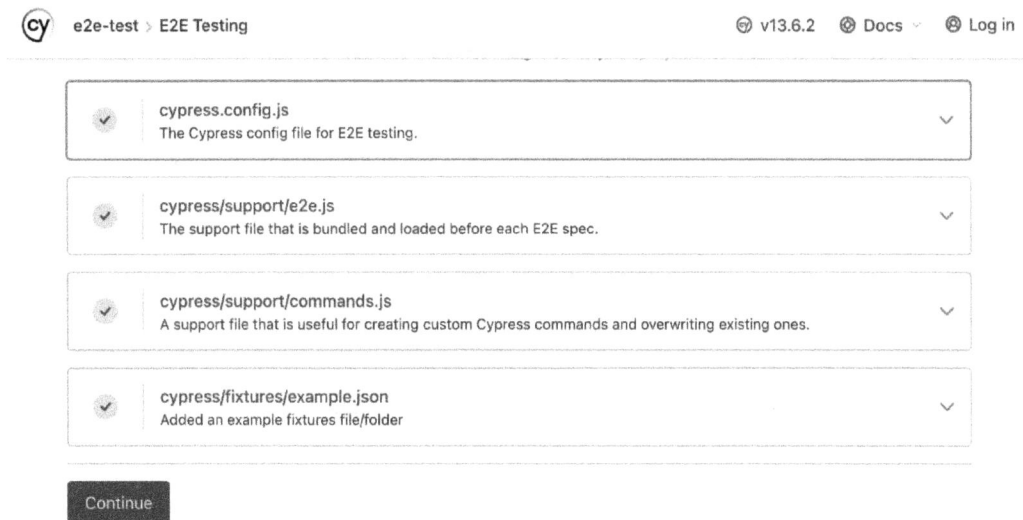

Figure 8.7: The Cypress wizard test file creation

The wizard has created four files. The first is the main configuration file that is hosted in the root of our project under the name `cypress.config.js`, which includes all the settings related to Cypress. The second and third are example files for support that are used to create reusable commands; commands are for more advanced users and are not covered in this book. The last file is a fixture example JSON, accessible at `cypress/fixtures/example.json`. Fixture files are used to save reusable information, such as text input or an API response.

If we continue our journey within the installation wizard, we will be presented with the browser selector screen. This screen will be displayed every time you run Cypress and allows you to select which browser you want to use.

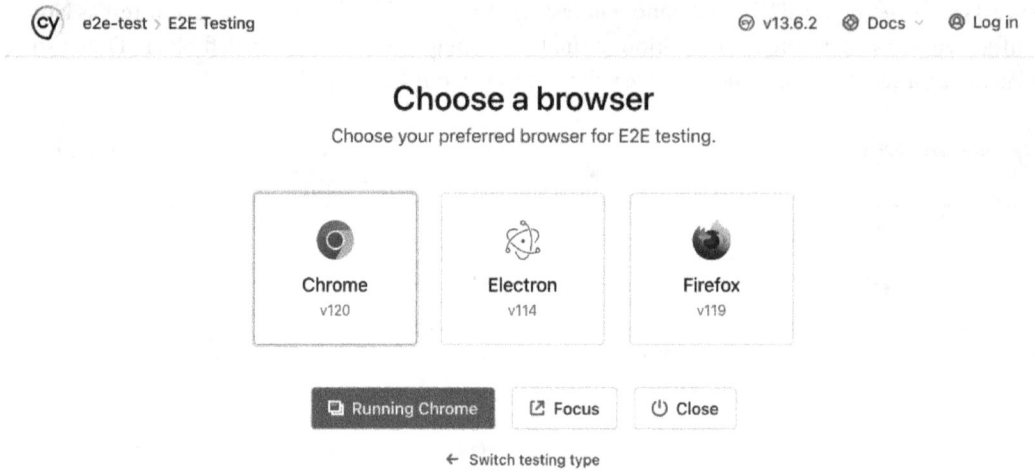

Figure 8.8: Cypress's Choose a browser screen

Let's open the test runner by clicking on Chrome.

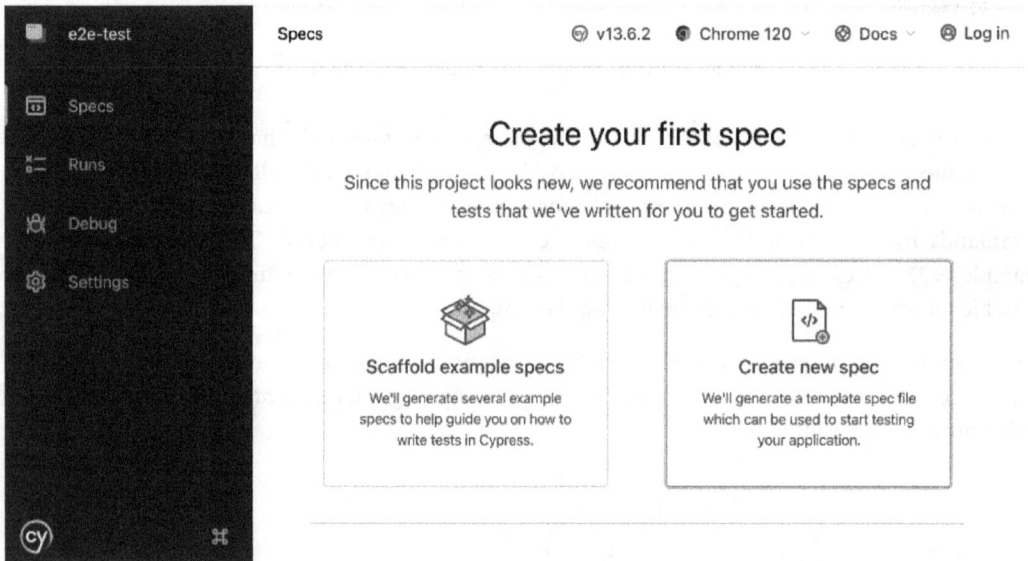

Figure 8.9: The first instance of Cypress's chrome test runner

Since we chose Chrome as our test runner, we will start in a new Chrome window. This runner will offer us the ability to create example tests or generate a test template.

We are going to stop at this point in the installation wizard; it is time to return to our Companion App and continue our journey there to learn more about E2E tests and Cypress.

> **Installing and reading the example tests**
>
> The tests provided by the "scaffold example specs" are very extensive and well-written. These sample tests provide you with an amazing insight into what E2E tests can achieve. Spend a few minutes installing and reading through these tests before moving on to the next chapter.

Learning the E2E tests' file format and file location

In the previous section, we learned how to install Cypress into our projects. In this section, we will learn the file location and format of these E2E test files before moving on to create our first test. Remember that we are now working within our Companion App, so you should go and open it with your IDE of choice.

The location of our E2E tests is defined within the Cypress configuration file. As we learned from the previous section, this is located at the root of our application and is called `cypress.config.js`. This file includes a couple of settings, one of which is called `specPattern`. This configuration setting informs Cypress where to find the tests:

```
specPattern: 'cypress/e2e/**/*.{cy,spec}.{js,jsx,ts,tsx}'
```

This pattern expects the E2E files to be located within the `cypress/e2e` folder and for the filename to end with a combination of `cy`, `spec` and `js`, `jsx`, `ts` and `tsx`.

So, a test file could be called `mytest.spec.ts` or `mytest.cy.js`, and if it is within the `cypress/e2e` folder, the Cypress test runner will be able to see and utilize it.

Let's go and create our first test, called `homepage`. Because I want to distinguish between unit tests and E2E tests, I will end this file with `cy.js` (the unit tests end in `spec.js`). The full location of the new file would be as follows:

```
Cypress/e2e/homepage.cy.js
```

Now that the file has been created, it is time to learn how to structure these files. As I previously mentioned, there are some similarities between E2E and unit tests' structures. In fact, both tests follow the same syntax, with a structure based on callbacks and methods called `describe` and `it`, respectively.

A Test scaffold should look like this:

```
describe('Homepage', () => {
  it('default journey', () => {});
});
```

As you can see, the code structure is noticeable, like the one offered by Vitest, with the only difference being that, in this case, we do not need to import the `describe` and `it` methods, as they are automatically imported for us.

Another small difference is the name of the test. When we defined a name in a unit test, it was a very important part of the test, with the well-defined `Given`, `When`, and `Then` methodology. In E2E tests, names are a little less important, mainly because these tests can have a large scope, such as a full check of the home page, and providing a single sentence to define what we are testing is not always possible.

Before we move forward, we are going to try and run E2E tests in our application and learn the steps required to do so.

In the previous section, we used `npx cypress open` to start Cypress. This can still be used, but our `package.json` has a couple of scripts ready for us to use that come with extra configurations, simplifying our development experience:

```
"test:e2e": "start-server-and-test preview http://localhost:4173
'cypress run --e2e'",
"test:e2e:dev": "start-server-and-test 'vite dev --port 4173' http://
localhost:4173 'cypress open --e2e'"
```

The application offers two scripts, `"test:e2e"` and `"test:e2e:dev"`. The first is used to run E2E on a preview build, while the second is used to run the test on a development build with **hot reload**. This will allow us to make modifications to our application while the E2E test runner is ready to run.

Try to run the development E2E test environment by running the following command in the terminal:

```
npm run test:e2e:dev
```

The result of this command should be the Cypress dashboard that we saw before. Just like before, you should click on **E2E tests** and choose **Chrome** as the browser to run our tests.

The result should be a Chrome browser, as shown in the following screenshot:

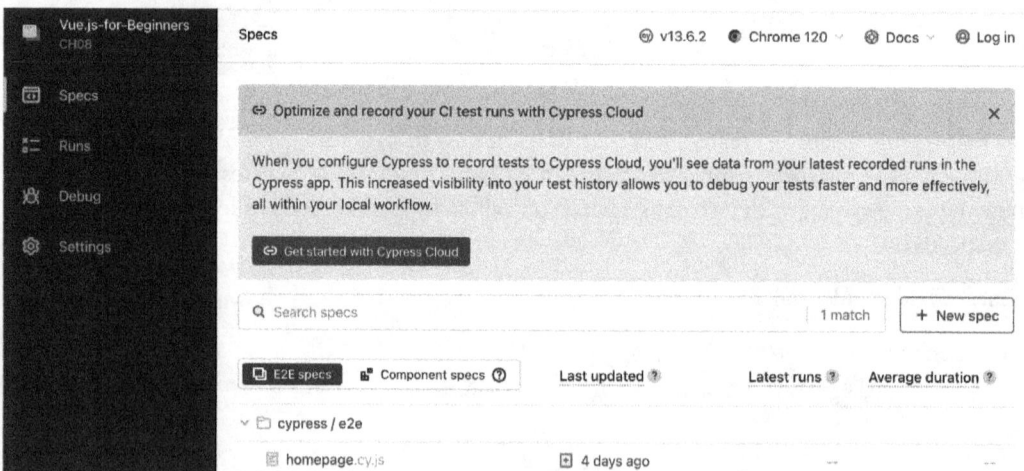

Figure 8.10: The Cypress test dashboard in Chrome

> **Browse around**
>
> Before continuing the chapter, you should spend a couple of minutes to familiarize yourself with the test runner. Knowing what it has to offer and how it works will be very beneficial for what you learn in the future and your usage of E2E tests.

To run our test, click on the test name, `homepage.cy.js`. This will load the Cypress test runner:

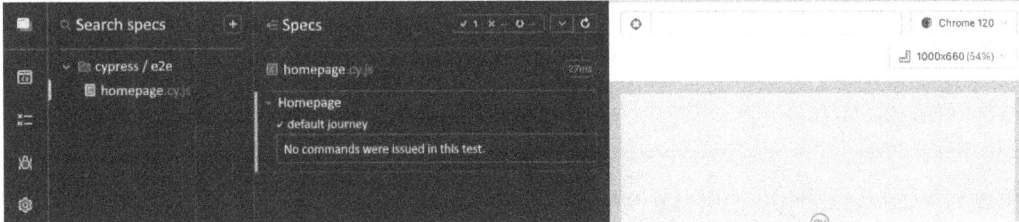

Figure 8.11: The Cypress test runner

Currently, the test runner is not very useful, as the only test we have is empty.

Now, it's time to move on to the next section, where we will write our first test in Cypress.

Writing your first E2E test

E2E tests are usually referred to as a **journey**. The name is derived from the fact that it encapsulates a specific user's journey. These could be the steps required to complete a **contact us** form, or the ones required to buy a product.

Due to the size of our current application, the journey will be quite small, but it can still be useful to ensure we build a stable application.

Our test will complete the following aspects:

- Accessing the site
- Displaying the app title
- Ensuring that posts are loaded
- Checking whether a post that has no comments, show the correct empty message
- Checking whether a post that has comments displays them correctly.

E2E tests are written sequentially, in the same way that you would complete the same journey manually. So, we are going to start our test by accessing the site on the main page. This is done by using the `cy.visit` command:

```
describe('Homepage', () => {
  it('default journey', () => {
```

```
    cy.visit('/');
  });
});
```

When using the `visit` command, you can pass any relative URL. In our scenario, we will just pass a forward slash, as we are going to visit the home page.

> **A base URL preset**
>
> Note that we were able to visit the home page by using a single forward slash because we have set a `baseUrl` preset within `cypress.config.js`. Without the `baseUrl` setting, you would have had to insert the full URL within the `cy.visit` command.

Next, we are going to check for the presence of the title `Companion App`. This is achieved using two new methods, `get` and `should`:

```
...
cy.visit('/');
cy.get('h1').should('be.visible');
...
```

The `get` method is just like `document.querySelector` and allows you to select a given element on a page. The `should` method allows you to set an expectation – that is, what defines a test as a pass or a fail. The `should` method accepts a parameter that is the logic by which we test our selector. These parameters are called **chainers** and have hundreds of different possibilities. The best way to learn all the available **chainers** is to either access the documentation (`https://docs.cypress.io/guides/references/assertions`) or use the IntelliSense IDE.

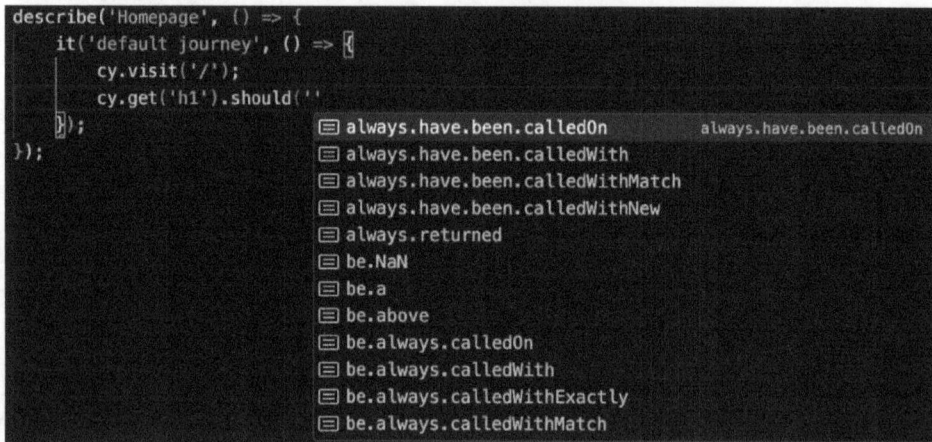

Figure 8.12: An IntelliSense pop-up of chainers

A list of chainers is displayed automatically within your IDE while you write your test expectation.

Whether you use the documentation or the IDE, as shown in *Figure 8.12*, all that really matters is that you familiarize yourself with these different chainers and understand what you can and cannot use.

In the preceding example, we just passed a single argument when calling the `should` method, but it can also accept two arguments. The second argument is used to pass a value when the condition defined in the first argument requires one. For example, we could be comparing the number of elements on screen with a variable, comparing two strings, or ensuring that the value returned by the API matches a specific object. In our current test, we just check that `<H1>` exists, but we are not really checking whether it is the correct heading, so we could change our implementation to use two arguments:

```
...
cy.visit('/');
cy.get('h1').should('contain.text', 'Companion App');
...
```

When using `contain.text`, we compare the `innerText` string of the element we selected with an arbitrary value.

The next step of our E2E test will be to ensure that the posts load successfully, by using the `get` method to fetch the element and the `have.length` chainer to ensure that the result value is what we expected:

```
...
cy.get('h1').should('contain.text', 'Companion App');
cy.get('.SocialPost').should('have.length', 5);
...
```

When the application loads correctly, it will load five posts, and to check this in our test, we select all the elements that have a class of `SocialPost` and compare their lengths, using the `should('have.length', value)` syntax.

Next, we will need to test the comment component. We will achieve this by clicking the **Show comment** button. If we check the HTML of the `SocialPost.vue` component, we would find out that the best way to find the button is to use a large selector, `.SocialPost .interactions button`. This selector is not an optimal solution, not only because it is very verbose, but also because it is too dependent on the structure of the component and, therefore, very fragile. To avoid using complex selection, we can add an E2E attribute. This is usually defined by adding a `data-cy` attribute to an element and using it for testing purposes (`cy` in the data stands for cypress).

Adding a data attribute to your code will make your test more robust, avoiding the creation of flaky tests associated with changes in CSS classes and HTML element structure.

Before continuing with our test, let's open `SocialPost.vue` and add the required attribute:

```
<TheButton
  @click="onShowCommentClick"
  value="Show comment"
```

```
    width="auto"
    theme="dark"
    data-cy="showCommentButton"
/>
```

data-cy is just a simple data attribute, but it is industry-standard and is used by many developers to provide a direct selector for E2E tests. We are now able to update our test to use the newly declared showCommentButton attributeto select our button:

```
cy.get('.SocialPost').should('have.length', 5);
cy.get('[data-cy="showCommentButton"]').first().click();
```

Since we have five posts on the page but just want to click on the first one, we will use a helper method called first that is used to retrieve just the first element returned by get(). Then, we end the chain by calling a click method to emulate a mouse click on the button. The result should be the comment component loaded.

If you check the browser window displaying our test runner, you should be able to see the comment component loaded in the first element.

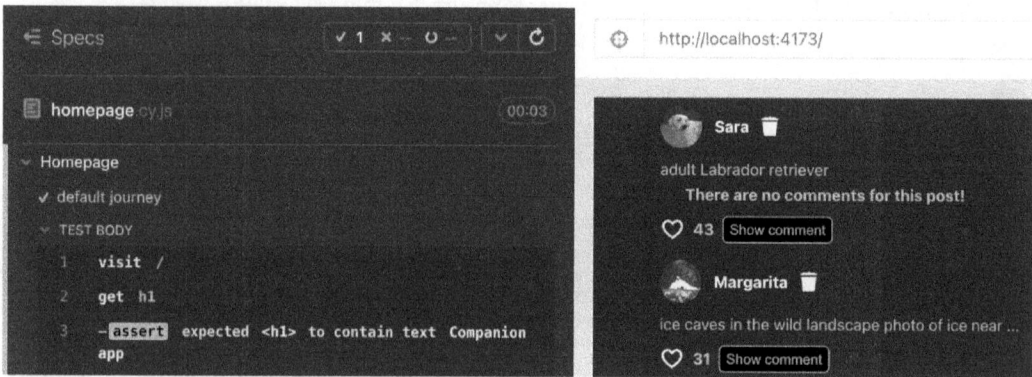

Figure 8.13: A Cypress test showing the Companion App with the comments loaded

Next, we are going to use the knowledge we learned in the last section to test the correct rendering of posts with comments. As shown in *Figure 8.13*, the first post has no comments. To better increase our coverage, we should also find a post with comments. A post with a comment is found in the last post (number 5), so we can use that to ensure that the comments are correctly displayed.

Based on what we have learned, we should be able to write a test that looks like this:

```
cy.get('[data-cy="showCommentButton"]').first().click();
cy.get('.SocialPostComments').should('contain.text', 'There are no
comments for this post!');
```

```
cy.get('[data-cy="showCommentButton"]').last().click();
cy.get('.SocialPostComments').last().should('not.contain.text', 'There
are no comments for this post!');
```

The code displays three commands. The first checks the content of the social post comment component. We are using the main class, `.SocialPostComments`, to select the component and the `contain.text` command. This command checks the `innerText` string of that selector and compares it with a string that is passed as an argument. In our case, we passed `There are no comments for this post!`. Then, we clicked the last post by replicating the command we previously used and replaced `first()` with `last()`. Lastly, we again replicated the command we recently used to check the content of the string, and we checked the opposite by adding the word `not` in front of the chainer. It is common when creating E2E tests to have commands that test something (e.g., a string or a number), and then a command that tests that that assertion does not happen again. Because of this recurring scenario, the E2E framework provides you the ability to add the word `not` in front of the chainer to check for the opposite. So, if this `should('contain.text', 'hello')` checks whether the element includes the word `hello`, then `should('not.contain.text', 'hello')` checks for the opposite – that is, to make sure that the element selected does not contain the word `hello`.

This is the end of our E2E journey. In the next section, we are going to introduce advanced techniques that have not yet been covered in this chapter but may turn out to be very useful for your future testing experience.

Introducing advanced testing techniques

In this closing section, we will introduce a couple of extra topics that are part of testing but have not been covered in this chapter. As I mentioned at the start of this chapter, testing is a very large subject, and this was just a quick introduction to ensure that you will know how to get started and create simple tests. However, it is by no means a complete guide, and further learning is required.

The aim of this section is to explain a few aspects of testing that you may find in your future experience when creating E2E testing or unit testing. Some of this applies to both E2E and unit testing, while others are unique to a single testing methodology.

Testing is a very interesting subject because what you learn about it is driven entirely by which you work. So, some people may have to learn how to write tests that heavily rely on external APIs and therefore focus on implementing stubs and mocks, while other developers may work on application that rely heavily on state management and may have to learn the ins and out of testing components that use state.

Let's dive in and share a few words on these three topics:

- Mocking
- Spying
- `shallowMount`

We will start with mocking.

Mocking

Mocking is a method used in both unit tests and E2E tests. The definition of mocking is as follows:

> *"Mocking is a process used in testing to isolate and focus on the code being tested and not on the behavior or state of external dependencies, by emulating a method or object."*

There are some cases, when writing a test, where you may want to avoid the use of real services (methods or objects). For example, you may want to avoid creating an order at every test run or having to use a paid API to fetch some dummy data. Another example would be mocking native APIs like fetch and IntersectionObserver.

To avoid these external dependencies, we can create a fake method or object that emulates the real third-party app or service. This is called **mocking**.

Mocking is an essential technique used to ensure that your tests are scoped and do not rely on external factors.

Spying

Spies are very similar to mocking; in fact, they give you the ability to analyze the use of a specific method. This information can then be used in testing expectations – for example, being able to assert that a method was called *X* number of times. The main difference from mocking is that spies do not actually change the original method and just listen to its usage.

Spying is very useful to ensure the correct execution of an application, without interfering with the actual methods. Adding a spy is very similar to using a proxy, where all requests for a spied method would first pass through the test framework.

You may need to spy on an action in a store to ensure that it is called during the execution of your test, spy on a log method to ensure that the correct values are passed to it, or keep an eye on a global method within the window object.

While mocking is done to prevent us from using external services, spying is more aimed at providing us with a tool to make the correct assumption and assertion within a test.

Just like mocking, we can spy on methods and modules using tools provided by testing frameworks.

shallowMount (unit tests only)

This last feature is available only in unit testing and, more precisely, to test components. In the *Unit testing with Vitest section*, we used the `mount` method to create an instance of our component, but there is another method available, called `shallowMount`, and in this section, we will explain the main difference between them.

Unit testing is all about speed, and it is essential when writing a unit test to always choose the faster approach that uses fewer resources and completes the test quicker. One of these economies can come from how we initialize our components.

When using `mount`, Vue Test Utils renders the component and any other component included in it. So, running mount on `app.vue` would render the complete application within it. Because a unit test is expected to be focused on a specific unit, you may need to render only the component you are testing and not its children.

To achieve this, we can use `shallowMount`.

`shallowMount` will render the component but then **stub** the children component by just rendering placeholder HTML. Doing so will reduce the resources needed by the test and make it more performant.

Choosing which one to use is due to your preferences and the overall architecture of your app. I personally prefer to use `mount` for most of the components, ensuring that they load correctly even with their dependencies, and I rely on `shallowMount` for complex components with several children.

Summary

In this chapter, we introduced the testing pyramid, covering the importance of tests and the different testing practices available within software development. We then moved on to unit tests and learned how to test our application using Vitest and Vue Test Utils. We then moved up to the testing pyramid and introduced E2E testing with Cypress. We created a small test that covered a simple user journey and learned a couple of techniques to select and test our application. Lastly, we closed the chapter with the introduction of future testing features that are part of the testing ecosystem and could be useful for future learning.

> **Your turn**
>
> Spend a couple of hours trying and testing more components to learn more about unit tests and expand the user journey of our E2E tests. Make sure to read the official documentation that includes all the available commands for both testing frameworks and is the best resource to find what you need.

In the next chapter, we are going to introduce two advanced techniques called `slot` and `Refs`.

Introduction to Advanced Vue.js Techniques – Slots, Lifecycle, and Template Refs

Up until now, we have learned about the basic features and techniques offered by Vue.js. Things such as properties and computed are the foundation of the Vue.js framework and will be used daily as you develop your next application with Vue.js. In this chapter, we are going to cover a few features that are what I call "advanced." This is not described as "advanced" because of its complexity, but in the fact that you will be less likely to use this daily. Features such as slots and Template Refs are used to solve specific use cases and are not expected to be encountered during your average task but are more likely to be used for specific situations that occur less frequently during a project. What you will learn in this chapter may not be used for quite some time, so it would be good to remember its existence and make sure you come back to it if the need for any of these features arises in your task.

In this chapter, we are going to add yet more features to our Companion App. We are first going to enhance our base button to introduce the notion of slots. Next, we will create a shared layout to be used in our static pages using named slots. Then, we will turn our attention to adding a new feature that will allow us to add new posts. While doing so, we are going to learn how to use Template Ref to access the **Document Object Model (DOM)** to autofocus the `CreatePost.vue` component. To conclude the chapter, we are going to build a completely new feature that will allow us to expand and collapse our sidebar. While doing so, we are not only going to iterate over the Vue.js lifecycles hooks but we are also using this opportunity to go over previously learned techniques such as methods, dynamic classes, and directives.

The chapter will be divided into the following sections:

- The power of slots
- Accessing components elements with Template Refs
- Deep diving into the lifecycle of a real app

By the end of this chapter, you will be able to use components that offer slots and develop components that expose one or multiple slots. You will also be able to access an element within a component with the use of Template Refs and finally be able to make better decisions when using lifecycles, to ensure that the application is free of bugs.

Technical requirements

In this chapter, the branch is called CH09. To pull this branch, run the following command or use your GUI of choice to support you in this operation:

```
git switch CH09.
```

The code files for the chapter can be found at https://github.com/PacktPublishing/Vue.js-3-for-Beginners.

The power of slots

In the first few chapters of this book, we have learned how properties and events can be used for a parent and a child to communicate with each other. This method of communication is not very flexible, as the only way for a component to expose information is by creating a new property.

In many cases, the rigidity that properties provide is precisely what we want to make sure that our component renders correctly, but there are times when more flexibility is needed, and this is where slots come in.

Let's consider our base button component. Its look and feel are defined by its properties and so is its value, but what would happen if we wanted to create a button with an icon before its value?

With our current knowledge of Vue.js, we would resort to creating a new prop of icon (prependIcon) that accepts an icon. Then, a further requirement may need us to add an icon after the value, so we would again resort to a new property of icon (appendIcon). Each additional request may result in a new property making our component very hard to maintain. Luckily, all the preceding requirements can be solved using slots.

Slots turn your component into a wrapper, allowing people to pass any arbitrary HTML or components into it. Slots are nothing new; in fact, native HTML offers a very similar functionality that you have been using automatically without knowing it. <div> is just a wrapper to other elements, <h1> can include text but also other elements in its content, and the list can go on.

The same functionality that a native button, heading, or span offers is also offered by Vue.js slots.

Let's consider how we would use a button in HTML. We would open the element with `<button>` and then we would add something within its content – this could be simple text, another element, or both. Finally, we would close the element with `</button>`, just like this:

```
<button>
  <icon src="myIcon" />
  My button with icon
</button>
```

Well, the `My button` text and the `<icon>` element displayed in the previous code snippets are what we refer to in Vue.js as **slot content**. As previously mentioned, using properties is a great way to ensure that the values passed to the component are of a certain type, but this advantage can easily turn into a disadvantage that makes the component quite rigid.

Let's see how to use slots by modifying `TheButton.vue`. We are going to remove the property called value and replace it with a slot so that our button will act as a native HTML button.

Let's modify our component:

```
<template>
  <button :class="theme">
    <slot></slot>
  </button>
</template>
<script setup>
defineProps({
  value: {
    type: [String, Number],
    required: true
  },
  width: {
    type: String,
    default: "100px"
  },
  theme: {
    type: String,
    default: "light",
    validator: (value) => ["light", "dark"].includes(value)
  }
})
</script>
```

First, we removed the property called `value` from the `defineProps` object, and then we added a slot in our component template by adding a Vue.js element called `<slot>`. This element does not need to be imported as it is accessible globally.

Now, we need to find all the occurrences of `TheButton` and replace the syntax from the previous one that used the property value:

```
<TheButton value="Example value" />
```

We need to replace this with the new one that uses a slot:

```
<TheButton>Example value</TheButton>
```

The preceding change needs to happen in `CreatePost.vue`, `SocialPost.vue`, and `Sidebar.vue`.

> **Slots are more than just text**
>
> Remember that now our component value not only accepts text, but it can also accept other HTML elements and Vue components.

Now that we know the basics of slots, let's move on and learn about them in more detail, starting with the slot default value.

Adding default values to a slot

Vue.js slots not only allow us to replicate native functionalities, but they also offer some extra features, one of which is the ability to add default values to our slots. This could either be a dialog heading or the text shown in a form submit button; default values can help you keep your code clean.

Adding default values for your slot is very simple; all you need to do is add the value that you want directly within the `<slot>` declaration. This value is going to be used if no other alternative is passed to the component, otherwise, it is going to be removed and overridden from the slot received.

Let's see a quick example to understand the declaration and usage of default slots.

First, we are going to add a default text of `Click Me` to `TheButton.vue`:

```
<template>
  <button :class="theme">
    <slot>Click Me</slot>
  </button>
</template>
```

Now that we have added a default value, our button will automatically show the new text, in the case in which no content is passed.

To see the default slot in action, we would have to call our button without any slot content, so the code would look like this:

```
<TheButton></TheButton>
```

This will render our default text.

Figure 9.1: Default button with a value of "Click me"

The text is there just as a fallback; in fact, adding a value within the slot will override the text:

```
<TheButton>Show this</TheButton>
```

Just as we saw before, adding text in the slot will display it, just as you would expect in a native HTML element:

Figure 9.2: Default button with a value of "Show this"

Slot versus property

Before we learn one more feature about slots, I want to write a few words to clarify why we need two methods (slots and properties) to achieve a very similar result.

Slots and properties solve two different use cases, and both have advantages and disadvantages:

- **Slots**: These are used mainly for components that require the values to be flexible. These are components that just provide a structure, for example, a dialog or end elements such as buttons or headings.

- **Properties**: These are used for elements that require fine control of what is being passed to them. This could either be due to some style that would break with different content or because of the need to validate or format the value received.

We can use native HTML elements to emphasize the differences we just defined. In favor of slots, we have headings tags such as `<h1>` and `<h2>`. These are very generic elements that provide styles, mainly in the form of font size and spacing to its content. This element requires flexibility, and the use of slots is perfect for it. In fact, native heading elements wrap their content just like slots.

On the other hand, we have the `<input>` element to prove the usefulness of attribute values (properties in Vue.js). The input element requires a specific attribute called `value`. The value offered by the input field only accepts a string or a number as its value. Offering a syntax that accepts an attribute, `<input value="text" />`, ensures that the value passed to the input field is validated.

Provide multiple slots with named slots

As you have probably noticed, slots have lots of useful features and can be very handy for constructing complex components. In the previous section, we have seen how we can define a single slot, but Vue. js offers more than that.

In fact, components in Vue.js can define more than one slot at a time. This is achieved using a functionality called **named slots**.

A common use of named slots is to define layouts. You can define a layout that accepts a sidebar, main content, and footer and then allow users to pass all the content that they see fit within the individual section. Another good example is a main hero component that offers a heading and a subheading, or a dialog component that offers a title and its content.

Let's create a very basic layout for our static content. We are going to store our new component in a folder called `templates` and call it `StaticTemplate.vue`. The relative URL would be `src/components/templates/StaticTemplate.vue`:

```
<template>
  <h1></h1>
  <main></main>
  <footer></footer>
</template>
<style scoped>
h1, main, footer {
  grid-column-start: 1;
  grid-column-end: 3;
}
h1{
  align-items: center;
}
main {
  padding: 16px 32px;
}
footer {
  border-top: solid 1px lightgray;
}
</style>
```

Our layout is going to have some basic styles that are required to create the correct spacing and show the difference between the individual sections. The template also includes three different sections: a heading defined with `<h1>`, the main content defined with the `<main>` element, and finally, `<footer>`.

Let's now add the named slot to our templates. The syntax of the named slot is very similar to the normal slot, therefore defined using the `<slot>` element, but with the addition of a `name` attribute. Just to emphasize the previous learning material, we are going to provide a default value for the footer:

```
<template>
  <h1><slot name="heading"></slot></h1>
  <main><slot name="default"></slot></main>
  <footer>
    <slot name="footer">
    Copyright reserved to Vue.js for beginners
    </slot>
  </footer>
</template>
```

With the preceding code, the `StaticTemplate.vue` file can be called with the possibility to pass three different sections. Let's go over the `AboutView.vue` page within the `Views` folder and try to use this new layout.

The syntax for the named slot is a little different than the default slot, as we need to define the actual name of the section that we are defining. To use a named slot, you use the `<template #slotName></template>` syntax.

Let's try to apply this to our `AboutView` page:

```
<template>
<StaticTemplate>
  <template #heading>About Page</template>
  <template #default>
    This is my content and default slot
  <template>
</StaticTemplate>
</template>
<script setup>
import StaticTemplate from '../components/templates/staticTemplate.
vue'
</script>
```

First, we imported the new component in the script section and then we used the component just like we would use `<div>` or `<span>`. Last, we added content for our slot by defining a heading and default section. Because **About Page** does not require the overriding of the footer, we are leaving that out of our instance so that the default value can be rendered.

If we access **About Page** at "`http://localhost:5173/about`", we should see the following:

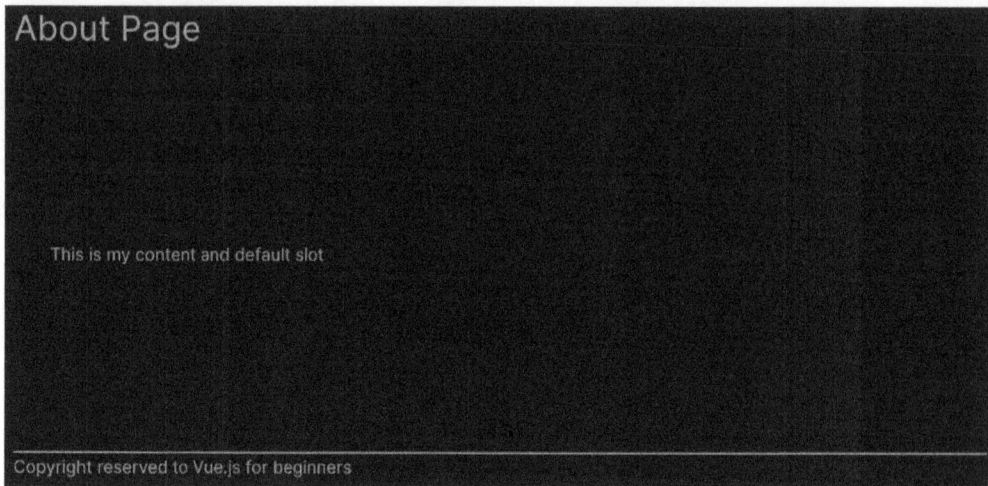

Figure 9.3: About Page using the new layout

The page is displaying all three sections as we expected.

> **The default slot does not require the <template #default> syntax**
>
> If we wanted to, we would have been able to call the default slot without the need to use `<template #default>`, but this is not a suggested approach as it would have produced a hard-to-read code with mixed methodology.

There is one more feature offered by Vue.js slot implementation called the **scoped slot**. This is a very advanced technique that allows you to define slots that expose the child's scope to its parents. Due to their complexity, these are not going to be tackled within this book.

Slots are a very powerful technique that helps us make our components easier to use, improve their readability, and offer enhanced flexibility. Just like every other feature, slots have their own advantages and disadvantages and should be used just when needed. The flexibility caused by the use of slots can produce unwanted side effects with the UI and it can be easier at times to control the layout of a component by imposing its values with the use of properties.

In this chapter so far, we have introduced the notion of slots by comparing them with existing features available within native HTML elements, and we then learned the syntax required to define and use default slots. After, we compared slots and properties and tried to distinguish the different scenarios in which one method may be preferred to another. Last, we learned how to define default values for a slot and the possibility of defining multiple slots with the named slot.

We are going to continue the chapter with another advanced topic, introducing how we can access children's components or HTML elements with the use of Template Ref.

Accessing components elements with Template Refs

In this section, we are going to improve the control of our components by learning how to access the DOM element.

When I first started to learn about Vue.js, I was astonished by how much could be accomplished without the need to access DOM elements. The Vue.js engine is structured in a way that we can accomplish all basic operations with the use of props, data, methods, and computed properties.

Just like everything else, there are times when we may need some extra control, and this can be achieved with the use of `Template Ref`. Using this feature exposes the DOM element in which the Ref was defined. This would be the same as using `querySelector` or `getElementById` offered by vanilla JavaScript.

Same Refs different usage

You may be wondering why we are learning about Refs if we already learned about it at the start of the book to define the private properties of a component. Well, this Refs is different. It is defined with the same syntax, but it is going to hold the value of an HTML element or component instead of primitive data such as strings and numbers.

Refs can be used for two different scenarios. The first is accessing a native element to access its native APIs; one such thing could be to focus an element or trigger native validation for an input field. The next is to access another Vue component or package to expose methods or their inner information. A common scenario for this is the need to set or fetch the value of a WYSIWYG (What you see is what you get) or code editor. The second scenario is not common practice and should be used just in case of emergency. The reason why accessing a child component using Template Ref is not suggested is that it will couple the two elements together and this should always be avoided in component-based architecture.

Focus an onMounted element

Let's start to play with Template Refs by learning how to fetch the reference of an input field and focus it when the page loads. The focus feature could have been achieved with the `autofocus` attribute, but we will develop the feature manually to showcase the use of Refs.

We are going to work in `CreatePost.vue` and autofocus the `textarea` used to write a new post.

We need to make three modifications to our component. First, we are going to define a `ref`, just like we learned throughout this book. Next, we are going to assign the Template Ref to `textarea` and we are going to access that element using the `onMounted` lifecycle and focus the component:

```
<template>
  <form>
    <h2>Create a Post</h2>
      <textarea rows="4" cols="20" ref="textareaRef">
```

```
      </textarea>
      <TheButton>Create</TheButton>
  </form>
</template>
<script setup>
  import { onMounted, ref } from 'vue';
  import TheButton from '../atoms/TheButton.vue';
  const textareaRef = ref(null);
  onMounted( () => {
    textareaRef.value.focus();
  });
</script>
```

Working with Refs follows a specific order; we first create the const textareaRef = ref(null) Ref and, at this point, the Ref is going to be null. We then assign the ref to a specific element within our component; in our scenario, we add it to textarea. The Ref is still empty because the component is not rendered, and the element does not exist on the page yet. Finally, we trigger our logic within the onMounted lifecycle. Because the onMounted lifecycle is triggered after the component is fully rendered, the Ref will be fully defined and ready to use.

Before we move on, I want to focus on three more parts of the code block. First is a reminder that Refs in Vue.js require .value to be appended to them to be able to access their internal value. This applies to both normal refs, such as strings and numbers, and Template Refs which we are covering in this section of the book. Second, I want to emphasize the fact a Tempalte Ref associated with an element (in our case, texareaRef) is going to be null until onMounted is triggered. This means that if we ever use it in a computed property or method, we need to ensure that we check for null values to avoid errors. The last is about the naming convention. It is important for the name of the Template Ref and the one used in the HTML element to match for it to function. It is important for the name of the Template Ref variables (const textareaRef) and the one used in the HTML attribute (ref="textareaRef") to match for it to function successfully.

> **Refs are just for exceptional cases**
>
> Vue.js provides us with mostly everything we need out of the box, so Refs should not be overused. The use of Refs should be saved for specific use cases and not become the norm. If you find yourself using this more than a handful of times in a project, you are probably misusing other Vue.js features.

Access from native validation

This section will focus on advanced topics; we are going to reiterate the topic we just introduced by implementing another example that uses Template Ref to access an element. We will continue to work on the code that handles the creation of a post by adding some validation using the native HTML validation offered by HTML form inputs.

We are going to continue our work in `CreatePost.vue`. First, we are going to add some specific validation within our `<textarea>` element:

```
<textarea
  rows="4"
  cols="20"
  ref="textareaRef"
  required="true"
  minlength="10"
></textarea>
```

We made `textarea` required and defined a minimum character length of `10`. Next, we are going to create a new ref, and this time, we will gain access to the main `<form>` element within the page:

```
<template>
<form ref="createPostForm">
  ...
</form>
</template>
<script setup>
import TheButton from '../atoms/TheButton.vue';
import { onMounted, ref } from 'vue';

const textareaRef = ref(null);
const createPostForm = ref(null);
```

Just like in our previous example, the creation of a Ref requires two steps. First, we add an attribute of Ref to an element, and then we create a constant that has a name matching the Ref.

In this last step, we are going to create a method that will be called when the form is submitted and use our newly created Template Ref to access the native form validation API:

```
<template>
  <form ref="createPostForm" @submit="createPost">
    <h2>Create a Post</h2>
    <textarea
      rows="4"
      cols="20"
      ref="textareaRef"
      required="true"
      minlength="10"
    ></textarea>
    <TheButton>Create Post</TheButton>
  </form>
</template>
```

```
<script setup>
import TheButton from '../atoms/TheButton.vue';
import { onMounted, ref } from 'vue';
const textareaRef = ref(null);
const createPostForm = ref(null);
const createPost = (event) => {
  event.preventDefault();
  if(createPostForm.value.reportValidity()){
    //code to create post
  };
}
...
```

To achieve this, we used our knowledge of events gained in *Chapter 6* and triggered a method called createPost on form submission. Then, we created a method that prevented the native submit from triggering and used the createPostForm Ref to check the validity of the form with createPostForm.value.reportValidity().

This method will return a Boolean (so true or false) depending on whether the form is valid and it will also display the native error message on the screen.

Let's start our application and try to trigger the form with an empty input:

Create a Post

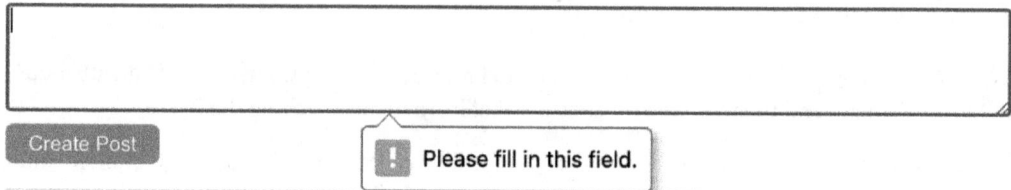

Figure 9.4: Native HTML validation error displayed on textarea

We are going to leave this form as it is even if it is not fully functional as the code required to submit the form will be completed in *Chapter 12*.

In this section, we have learned that Ref is not only used to define our component values but it is also used to hold values of HTML elements. We have applied this newly learned technique by developing the functionality and validation needed to create new posts in CreatePost.vue. Finally, we iterated on the flow in which Refs are defined and how it requires the mounted state to be completed for it to be accessible.

In the next section, we will add the ability to expand and minimize the sidebar of our application and iterate through the notion of the lifecycle.

Deep diving into the lifecycle of a real app

Lifecycles are not a new notion and have been mentioned once or twice throughout this book, but due to their importance, it is a good idea to iterate through them by adding further functionality to our Companion App.

In this section, we are going to add new functionality that will enable us to expand and minimize the left sidebar of our application and store the value in `localStorage` to ensure the user preferences are persistent on refresh. We are going to learn how having a better understanding of the lifecycle can improve the user experience provided by our application.

Before we jump into the code itself, let's think about what steps would be required to achieve this new functionality. Thinking over a problem helps you memorize all the different aspects of the Vue.js framework that we have learned and supports you in realizing any lack of knowledge or misunderstanding that you may have.

To complete the task of enabling the sidebar to collapse and expand, we would need to do the following:

1.   Add the UI required for the user to trigger this action.

2.   Add some style to improve the look and feel of the sidebar.

3.   Create a new value that will hold the state of the sidebar.

4.   Expand the sidebar to have different renderings for the different states.

5.   Add the logic to handle the sidebar state change and save the value.

6.   Finally, read the user preferences on load and apply them to the component.

The list may seem very long, but these are all small tasks that you should be able to accomplish without my aid. Let's start our work by working on the new UI changes.

Adding conditional rendering to the sidebar

Let's start by reiterating the Vue.js directive by using `v-if` and `v-else` to effectively display two different views. One will show the expanded sidebar, while the other will display a minimal layout with just the button required to expand the sidebar.

All the following developments will take place in the `SideBar.vue` file that holds the logic required for our sidebar. Before we make changes to its layout, we are going to import two new icons, `IconLeftArrow` and `IconRightArrow`:

```
<script setup>
import { ref } from 'vue';
import TheButton from '../atoms/TheButton.vue'
import IconLeftArrow from '../icons/IconLeftArrow.vue'
import IconRightArrow from '../icons/IconRightArrow.vue'
...
```

Next, we are going to add the two layouts using the conditional logic provided by `v-if`/`v-else`:

```
<template>
<aside>
  <template v-if="">
    <IconRightArrow />
  </template>
  <template v-else>
    <h2>Sidebar</h2>
    <IconLeftArrow />
    <TheButton>Create post</TheButton>
    <div>Current time: {{currentTime}}</div>
    <TheButton @click="onUpdateTimeClick">Update Time</TheButton>
  </template>
</aside>
</template>
```

`<aside>` includes two different blocks, both delineated by the `<template>` element. The first block just contains the right arrow icon and will be used to display our collapsed layout, while the other layout includes the existing code that was present in the component before we started the modification, with the addition of the left arrow icon.

If you have looked closely at the preceding code, you may realize that something is missing, or incomplete. In fact, a conditional statement such as an if/else requires a hypothesis and this is currently missing in our code: `<template v-if="" >`.

For the statement to work, we need to add a condition that will be evaluated by the Vue.js compiler to display the correct layout. We are going to create a `Ref` called `closed` that will hold a Boolean value. This variable will be used by our if/else statement to define which layout should be displayed.

To accomplish this, we first define the `ref` in our script section:

```
<script setup>
```

```
import { ref } from 'vue';
...
const currentTime = ref(new Date().toLocaleTimeString());
const closed = ref(false);
```

Next, we use this new variable within the `v-if` directive that we have previously defined:

```
<template>
<aside>
  <template v-if="closed">
    <IconRightArrow />
  </template>
```

Due to the value of `closed` being hardcoded to `false`, the sidebar will always display the expanded layout. In the next section, we will write the logic required to toggle the two layouts.

Writing the logic to handle the sidebar states

To be able to toggle between the two different views of our sidebar, we need to provide the user with the ability to change the value of `closed`. We will use the left arrow and right arrow icons added in the previous section to trigger a click event on user input. The event will initiate a method that we are going to call `toggleSidebar`.

To code this, we are first going to add the event handler to the two icons:

```
<template v-if="closed">
  <IconRightArrow @click="toggleSidebar" />
</template>
<template v-else>
  <h2>Sidebar</h2>
  <IconLeftArrow @click="toggleSidebar" />
  <TheButton>Create post</TheButton>
...
```

Then, we declare the `toggleSidebar` method in the `<script>` section of our component:

```
const toggleSidebar = () => {
  closed.value = !closed.value;
}
```

The method accepts no argument and just redeclares the value of `closed` by making it equal to the opposite of the current value of `closed`. This generates a logic that toggles a Boolean value between `true` and `false`.

If you run the application now, you should be able to toggle between two layouts. The next step involves adding some styles to the collapsed layout that is currently still showing a full-size sidebar.

Let's add some styles to our component:

```
<template>
<aside :class="{ 'sidebar__closed': closed}">
  <template v-if="closed">
    <IconRightArrow class="sidebar__icon" @click="toggleSidebar" />
  </template>
  <template v-else>
    <h2>Sidebar</h2>
    <IconLeftArrow class="sidebar__icon" @click="toggleSidebar" />
    <TheButton>Create post</TheButton>
    <div>Current time: {{currentTime}}</div>
```

```
    <TheButton @click="onUpdateTimeClick">Update Time</TheButton>
  </template>
</aside>
</template>
<script setup>...</script>
<style scoped>
aside {
  display: flex;
  flex-direction: column;
  position: relative;
  &.sidebar__closed{
    width: 40px;
  }
  .sidebar__icon{
    position: absolute;
    right: 12px;
    top: 22px;
    cursor: pointer;
  }
}
</style>
```

To better style our two layouts, we declared two classes. The first is sidebar__closed, which is used to reduce the width of our sidebar. The second is sidebar__icon, which is used to define the size and position of the arrows.

sidebar__icon is applied to both icons, while sidebar__closed is just assigned to <aside> if the value of closed is true. To do so, we used the :class="{ 'sidebar__closed': closed}" syntax. This syntax is useful as it allows you to create complex styles with the ability to easily apply classes when a specific condition is met.

At this stage, the sidebar not only works but it is also styled properly with a collapsed and expanded layout. What is left to do is to make the data *persistent*. In development, we describe data as persistent when its value remains consistent, even after a browser refresh.

Figure 9.5: Expanded sidebar

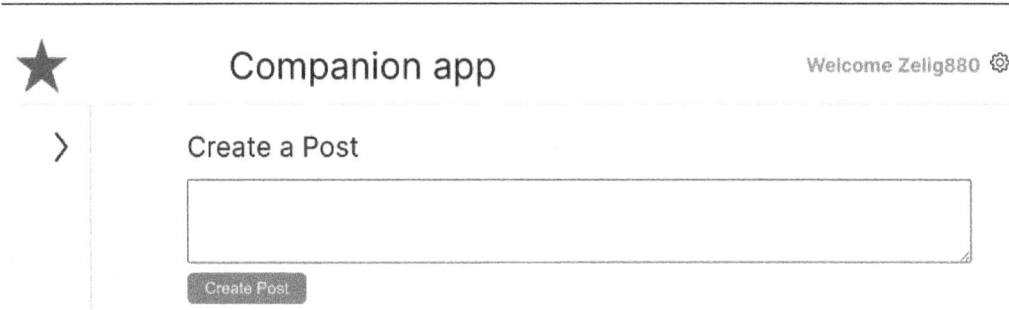

Figure 9.6: Collapsed sidebar

Now that the sidebar's look has been improved, it is time to make its value persist on refresh.

Save and read the user preference in localStorage

At this stage, even if the sidebar logic is fully working, its data is not persistent yet. In fact, if you set the sidebar to its collapsed state and refresh the page, you will see that it will go back to the expanded view that was defined as its default state (the value that we used to initialize the closed ref was false).

To enable persistence, we are going to use localStorage to save our value and re-read it on page load.

Before we go through the code, we should try and define what the best method is for us to achieve this. In fact, this section was named after the Vue.js lifecycle, but up to this stage, we have not used them yet.

> **Why is the use of the correct lifecycle important?**
>
> Before moving forward, spend a few minutes to try and understand what the consequences would be of using the incorrect lifecycle to load our data from localStorage. Think about the different lifecycles, when they are triggered, and what could be the effect on the application.

As we learned in *Chapter 2*, there are different lifecycles that support different use cases. In our scenario, we are planning to use a lifecycle to read the value of our closed variable and apply it to the component. When making an operation like this one, you usually should ask yourself a couple of questions. The first is whether the data is asynchronous, and the second is whether the data is required before the application is rendered on screen.

Each lifecycle happens at different stages of the component life. Some such as beforeCreate is triggered before the component is even created, while others trigger after a component is fully mounted within the DOM like onMounted, and it is important to select the correct one for our specific scenario.

In our case, the data is fetched from localStorage and is a synchronous operation; it is needed before the component is fully rendered or screened, also referred to as "before mounted.".

The lifecycle that best fits our needs is onBeforeMount. This will trigger just before the component is rendered but after all the methods and Refs have been initialized.

Let's add this logic to our component:

```
<script setup>
import { ref, onBeforeMount } from 'vue';
import TheButton from '../atoms/TheButton.vue'
import IconLeftArrow from '../icons/IconLeftArrow.vue'
import IconRightArrow from '../icons/IconRightArrow.vue'
const currentTime = ref(new Date().toLocaleTimeString());
const closed = ref(false);
const toggleSidebar = () => {
  closed.value = !closed.value;
  window.localStorage.setItem("sidebar", closed.value);
}
...
onBeforeMount( () => {
  const sidebarState = window.localStorage.getItem("sidebar");
  closed.value = sidebarState === "true";
});
</script>
```

To achieve persistence, we first imported the onBeforeMount method from the Vue package and then we saved the value of the closed Ref every time the toggleSidebar method was triggered. To achieve this, we used the localStorage.setItem method that is available in the window object. Lastly, we used onBeforeMount to read the value from localStorage and then assign it to the closed ref.

At this stage, our application will not only allow the user to toggle the sidebar, but its value will also be persistent on refresh.

Before we complete this chapter, I want to focus on why using the correct lifecycle is important. In fact, if we had used another lifecycle such as onMounted, the sidebar would have been fully rendered (incorrectly) before the value of localStorage would have been read and applied to the app. The main issue with bugs such as this one is that in development they may not reproduce or may be very hard to see.

It is very important when creating code that is going to alter the view of the component to ensure that you have used the correct lifecycle and, if working with asynchronous data, that the correct loading is defined, or the promise awaited before the rest of the component executes. This practice may seem hard to comprehend at first, but code practice and mistakes will help you improve your understanding of a Vue.js component and improve your skills.

Summary

In this chapter, we have covered a couple of advanced topics such as slots, lifecycles, and refs. The aim of this chapter was not to provide you with all the information that you may need on these topics but to introduce you to these points so that you can practice them in your next development and continue to learn as you expand your knowledge of Vue.js.

We have learned how to use slots to expand the flexibility of our components. Slots and named slots can be used for simple cases, such as `<button>`, style elements, such as `<div>`, or for more advanced techniques, such as defining a page layout with different areas.

We then moved on to Template `Ref`, a topic that we had partly introduced in an earlier chapter. We learned how Template `Ref` can be used to access a component DOM element. This is defined as an advanced technique because, with all the available features offered by Vue.js, it is uncommon that you would need to use Template `Ref` this way.

Lastly, we went over lifecycles again. Vue.js lifecycles are very important, and they require lots of practice to help you understand their usage and, more importantly, their order of execution. We have added an extra feature to our Companion App so that we can understand one of its use cases and think over the possible result that a different lifecycle would have produced.

In the next chapter, we are going to learn how to define multiple routes with **vue-router**. Defining multiple pages is a required step for most applications and **vue-router** offers a very simple syntax that will help us achieve this functionality within our Companion App.

10

Handling Routing with Vue Router

Single-page applications (**SPAs**), such as the one offered by Vue.js, provide an architecture built upon a single page. This methodology prevents the page from fully reloading and offers an improved user experience.

As your application grows, you will find the need to create different views in your application. Even if the term *SPA* can lead to the understanding that your application will be built on a single page, the truth is far from it.

Most frameworks, including Vue.js, offer packages aimed at recreating the routing system available in other frameworks such as PHP, .NET, and more. The routing functionality offered by SPA frameworks provides the best of both worlds. It offers developers a complete routing toolkit, while still providing the same user experience expected by SPA frameworks. The router package used in Vue.js is called vue-router.

In this chapter, you will learn how to use routing in your Vue.js application. At the end of the chapter, you will have a good understanding of vue-router and its configuration, and how routes are defined and used. You will be able to create basic, dynamic, and nested routes and navigate to them using different methodologies. Lastly, you will learn how to improve your user experience with the use of `redirect` and aliases.

We are first going to learn about vue-router by covering its configuration. We will then learn how to implement our first route and route navigation by creating a couple of static pages. Next, we will introduce dynamic routes by defining a user profile page. Then, we will add another level of navigation by splitting the user view into a user profile and user posts using nested routes. Finally, we will complete the chapter by familiarizing ourselves with `redirect` and `alias`.

This chapter is going to be broken down into the following sections:

- Introducing vue-router
- Navigating between routes
- Dynamic route matching
- Nested routes
- Reusing routes with `alias` and `redirect`

Technical requirements

In this chapter, the branch is called `CH10`. To pull this branch, run the following command or use your GUI of choice to support you in this operation: `git switch CH10`.

The code files for the chapter can be found at `https://github.com/PacktPublishing/Vue.js-3-for-Beginners`.

Introducing vue-router

vue-router is the official router package built and maintained by the Vue.js core team and community members. Just like other packages that we have introduced so far in the Companion App, also vue-router was automatically set up for us when we initialized our application using Vite.

In this section, we will learn about the file structure and configuration required for vue-router and introduce some of the syntax used when working with routes.

vue-router offers a standard set of functionalities that are expected from a router. So, if you have previously worked with routers in other languages, most of what we will cover will sound familiar, but it is still worth a read as the syntax may be different.

Learning about vue-router configuration

Let's first start by learning how to best configure the router in an application. In fact, even if vue-router is usually preset by tools such as **createVue** and **Vite**, it is important to understand how it is set behind the scenes.

The configuration required for the plugin to work is stored in a file commonly named `router.js` or an `index.js` file within a folder called `router`.

In our case, the file is stored within the `router` folder.

> **All plugins need to be registered in main.js**
>
> All plugins must be registered in `main.js` before they can work. So, if you ever want to find the configuration file or information for a plugin that is currently loaded in your application, you can open the `main.js` file and search for the `app.use(pluginName)` syntax.

Let's look at the syntax you might find in this `index.js` file:

```
import { createRouter, createWebHistory } from 'vue-router'
import HomeView from '../views/HomeView.vue';
const router = createRouter({
  history: createWebHistory(import.meta.env.BASE_URL),
  routes: [{
    path: '/',
    name: 'home',
    component: HomeView
  }]
})
export default router
```

Let's go over the important points of the preceding configuration. First, we can talk about the `createRouter` method. This method provided by the vue-router package creates an instance of the router that can be attached to a Vue application. This method expects an object including the router configuration.

Next, it is time to look at the first entry of the configuration object, `history`. The `history` property defines how your application will navigate between the different pages. In standard apps, this is usually achieved by changing the website URL to the desired page – for example, by appending `/team` to the URL to visit the Team page. Let's look at a couple of different configurations that can be used to set our `history` property.

Hash mode

This method is achieved by using `createWebHashHistory`. This method is provided by the vue-router package, and it is the easiest one to implement as it will not require any server-side configuration. When using hash mode, an actual hash (#) is going to be added between the base URL and our routes. With this configuration, accessing the Team page can be done by visiting `www.mywebsite.com#team`.

This method can impact your SEO negatively, so if your application is accessed publicly, you should invest time and set up the next available method, web history mode.

HTML5/web history mode

HTML5 mode can be configured by using the `createWebHistory` method. With this history mode, our website will act as a standard site with its route being served directly after the website URL (e.g., `www.mywebsite.com/team`).

Because SPA websites are built on a single page (hence the name), they can just serve the application from one single endpoint (the website base URL). So, deploying our site and trying to access the Team page directly would result in a `404` page (not found).

Solving this issue is a trivial task in today's hosting site, as all that is needed is a catch-all rule that ensures site navigation is funneled to the SPA entry point. If you wish to use this method, a little googling will help you find the instructions you need to properly set this in your hosting provider.

In our case, we are using web history. This is my default history setting, not only because it improves SEO but also because it has been the normal way to navigate our sites for years and I like to keep things consistent.

Finally, it is time to cover the last entry of our configuration: routes.

Defining routes

We have reached the most important part of our router configuration, the actual routes. The word *route* defines the ability of our application to direct a user to a specific page. So, when declaring routes, we define the pages a user can visit on our site.

To declare a route, we require two pieces of information, `path` and `component`, but I usually prefer to always include a third one called `name`.

The `path` attribute is used to define the URL that needs to be accessed for this route to be loaded. So, a path of `/` will be delivered if the user navigates to the base of your site, while a path of `/team` would be accessible on `www.mysite.com/team`.

The `component` attribute is the Vue component expected to be loaded when this route is accessed. Lastly, we have `name`. Adding this parameter to all our routes is good practice as `name` is used to navigate a route programmatically. We will cover this in more detail later as we will learn how to navigate within our application.

Now that we know all the different aspects of a route, let's try to decode what route was declared in the previously shared code snippet. The snippets show a route directing the user to the base path of our site (`path` is `/`), with the `name` value of `home`, which will load a `HomeView.vue` component.

Until now, we have learned how to configure a Vue instance, but there is one more step needed before our routes can function properly: adding `RouterView` to our application.

We have defined which component to load for a given URL, but we have yet to tell our Vue application where to load this component.

The way the router works is by replacing the content of our application every time a user navigates to a different page. So, in very simple terms, a router can be defined as a huge `if/else` statement that renders the component depending on the URL.

To allow the router to work correctly, we are going to add a component called `<RouterView>` to the main entry point of our application, the `App.vue` file:

```
<script setup>
  import { RouterView } from 'vue-router'
```

```
</script>
<template>
  <RouterView />
</template>
```

From this stage onward, vue-router is going to take ownership of what is displayed on the screen using the route definition we defined in our configuration.

Creating our first view

Let's try to add a new router navigation for a static page called **Privacy**. This is just going to include some dummy text for now.

To add a new page, we need two steps: a route defined in our `routes` array and a component that will be loaded when that route is accessed.

Components that are used as routes are stored in a folder called `views`. If we access this folder, we see that we currently have two views set, `Home` and `About`.

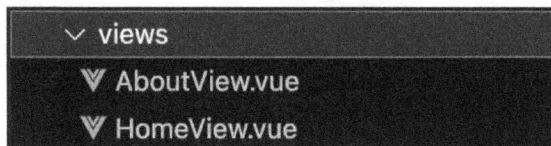

Figure 10.1: Folder tree showing the content of the views folder

We are going to add a new file to our folder called `PrivacyView.vue`. It is common to match the filename with the route's name. Because this file will be static, we will re-use the layout defined in *Chapter 9* when we learned about slots.

The layout used for the static page defined as `StaticTemplate.vue` accepts three different named slots: a heading, a footer, and a default template used for its main content. Our `PrivacyView.Vue` file content should be defined like this:

```
<template>
  <StaticTemplate>
    <template #heading>Privacy Page</template>
    <template #default>
      This is the content of my privacy page
    </template>
  </StaticTemplate>
</template>
<script setup>
import StaticTemplate from '../components/templates/StaticTemplate.vue'
</script>
```

You may have noticed that the file content just defines two slots (`heading` and `default`) and is missing the footer. This has been done on purpose as the footer has a default value that we want to display.

Before our new component can be visited and displayed we need to add it to our route. Let's do this by going back to `index.js` within the `router` folder and adding the route for our Privacy page within the `routes` array:

```
import { createRouter, createWebHistory } from 'vue-router'
import HomeView from '../views/HomeView.vue'
import AboutView from '../views/AboutView.vue'
import PrivacyView from '../views/PrivacyView.vue'

const router = createRouter({
  history: createWebHistory(import.meta.env.BASE_URL),
  routes: [
    {
      path: '/',
      name: 'home',
      component: HomeView
    },
    {
      path: '/about',
      name: 'about',
      component: AboutView
    },
    {
      path: '/privacy',
      name: 'privacy',
      component: PrivacyView
    }
  ]
})
export default router
```

Our new page views are first imported at the top of the page as a normal Vue component and then assigned to the new route. Just as we mentioned before, routes require three values. First, we set the `path` value as `/privacy`, then we defined a `name` value of `privacy` for future programmatic navigation, and lastly, we assigned the imported `PrivacyView` component to it.

After these changes, we should be able to see our page by accessing `http://localhost:5173/privacy`:

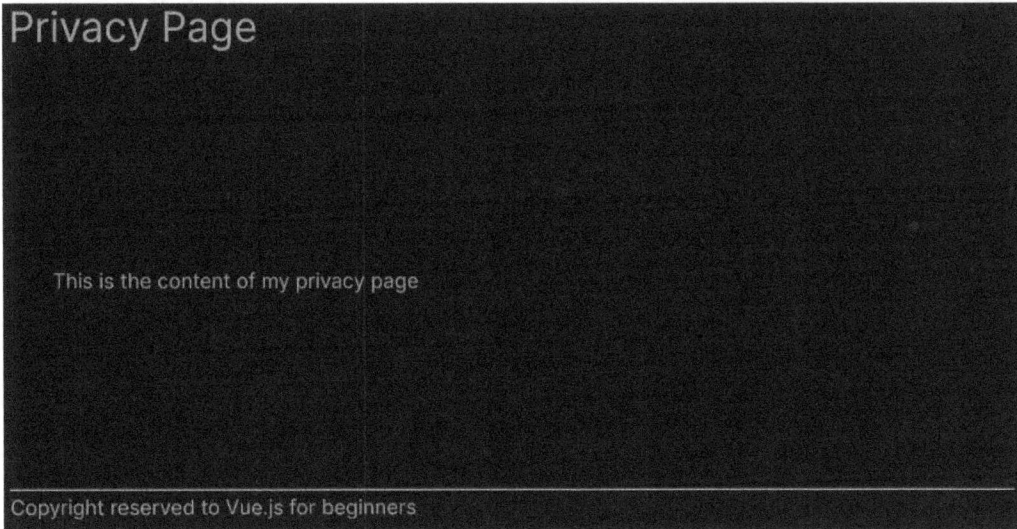

Privacy Page

This is the content of my privacy page

Copyright reserved to Vue.js for beginners

Figure 10.2: Screenshot of the Privacy page

In this section, we have learned how to configure vue-router, introduced routes and defined how they are used by the application to render pages, and finally, talked about `<RouterView>` and how vue-router uses it to display the correct page to our visitors.

In the next section, we are going to learn how to navigate between different routes.

Navigating between routes

Until now, we have learned how to create our routes and how to navigate through them by loading the URL directly in the browser. In this section, we will learn how to navigate between different routes directly in the code base.

It is true that we could define our navigation using a simple `<a>` tag, but that will force the app to fully reload on any navigation and so will go against the whole architecture of a SPA, which offers a "reload-free" experience.

To solve this issue, vue-router offers components and methods that will handle navigations without reloading the page.

While using vue-router, navigation can be achieved in two different ways. One uses a component called `<router-link>` while the other is triggered programmatically using `router.push()`. Let's see both methods in action and learn when to use them.

Using the <router-link> component

Using the <router-link> component to navigate within your application is a simple task, as it uses the same syntax offered by the native HTML <a> element. Behind the scenes, <router-link> is just an anchor tag with added functionality that prevents the app from fully reloading while navigating.

Our Companion App has three different pages, but there is nowhere to access them unless we enter the URL directly. Let's go and fix this now by adding two links in our sidebar, one for each static page we have created so far.

As a reminder, the main page is HomePage.vue, but the actual sidebar is just a child component of this page and can be found in the organisms folder under the name SideBar.vue. Now that we have located the file, it is time to add our router links.

First, we import the component from vue-router in the script section of the component:

```
import { RouterLink } from 'vue-router';
```

Next, we add the links in our sidebar, directly after the Update Time button:

```
<TheButton @click="onUpdateTimeClick">
  Update Time
</TheButton>
<router-link to="privacy">Privacy</router-link>
<router-link to="about">About</router-link>
```

The router-link component accepts a property called to. This property can accept the URL that the browser needs to navigate to or a route object, which we will cover later in the chapter.

In our example, we have specified the value of the to property to be equal to privacy for the link directing to the **Privacy** page and about for the page navigating to the **About** page.

After these two changes, the application will have two new links in the sidebar.

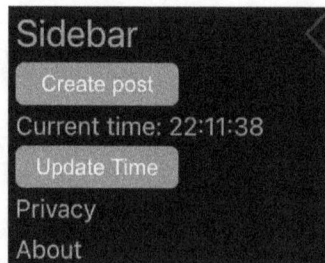

Figure 10.3: Companion app sidebar with the router links

Because router links are just simple <a> elements, they inherit the style of the anchor that was defined by the project template.

Programmatic navigation

For most cases, navigating using the router link component is all you need, but there are some situations in which you may find it beneficial to navigate through the app directly in the code.

There is no real difference in navigating using the `router-link` component or programmatically, and both methods are offered to ensure that your code can be clean and easy to read.

Programmatic navigation is useful when attached to a piece of logic. Being able to trigger the navigation manually helps you take control of the flow of the data, ensuring that your code is well written and that the user receives the best experience.

To be able to navigate to a different page, we can use a method offered by the `router` object called `push`. This method, just like `<router-link>`, can accept the URL or a `route` object.

We are going to add another link, but this time, a simple anchor element, and trigger the navigation using the `onclick` event.

First, we are going to add an import from the vue-router package called `useRouter`:

```
import { RouterLink, useRouter } from 'vue-router';
```

Then, we use this method to access the `router` instance. This just needs to be done once per file:

```
const router = useRouter();
```

Next, we are going to create a method that will handle the `click` event of our anchor:

```
const navigateToPrivacy = (event) => {
  event.preventDefault();
  console.log("Run a side effect");
  router.push("privacy");
}
```

Triggering navigation programmatically allows us to trigger side effects that would not be possible if using a `<router-link>` component, – for example, conditionally navigate a user to different pages depending on a value returned from the server.

Lastly, we are going to create a new anchor in the HTML section of our component. This element is going to use the `navigateToPrivacy` method:

```
<a @click="navigateToPrivacy">Programmatic to privacy</a>
```

We have now completed the first part of this chapter. At this stage, you have learned the basics of vue-router and gained an understanding of how to define routes, how they work, and how to navigate between the different pages.

In the next sections, we will learn about more advanced topics such as nested routes and dynamic route matching.

Dynamic route matching

If you are tasked to build a simple portfolio site, basic routes are going to be just fine. However, as you start to work on more complex sites such as blogs, you will need more complex routing. In this section, we are going to learn about dynamic route matching. Dynamic route matching is used to define routes that require one or more parameters to be dynamic.

All the routes that we have defined until now were static and did not change. So, we had the /about endpoint that would render the **About** page and the / path that would welcome the home page of our site. What would happen if we wanted to develop a blog post with our current knowledge? After our first blog post, we would write a route, /blog/1, then after the second, we would have to create another route, /blog/2, and so on.

Defining routes statically like this is not an ideal solution, and this is where dynamic route matching comes into place. Dynamic route matching allows us to create a pattern, for example, blog/:blogId, and then let the application render a specific page that uses the defined argument to render another specific page.

To learn about this great feature, we are going to enhance our Companion App by adding a new feature that will allow us to open a specific user profile.

The requirements for this task are the following:

- We are going to add the ability for the user to open a user profile page from the main view of the application
- We are going to create a new route that will display a user
- We are going to create a new component that will display the user with all its information

We are going to develop this backward, starting by creating the user-specific page.

Creating a user profile page

Our first task requires us to create a new page that will later be used by our route as the content of our user page. Just like all other pages, this will also live in a folder called views and will be called UserView.vue.

In this component, we are going to first load a specific user profile and then display its information on the screen. To ease development, we are going to hardcode userId so that we can see the correct information on screen for now, and then remove this later after our dynamic route is fully set.

Just like a simple component, we will need to define three different parts: the HTML, the logic required to fetch the correct information, and the styles.

To help you prepare for the real world, you could go a step further than this book and learn what information we would need from the API to develop the page. When developing from an external source, such as an API or a Content Management System (**CMS**), it is important to be able to fetch the structure of the object. In our specific scenario, we will use the `user` object, but what will that include? Would it have a name? Would it have a date of birth, and if so, what would the format be?

All these questions are usually answered by the **model** of the `user` object, which is usually available directly from API and CMS documentation. In our case, since we are using `dummyapi.com`, the information about the API models can be found here: `https://dummyapi.io/docs/models`.

Within the model, we can find our specific one, called **User Full**:

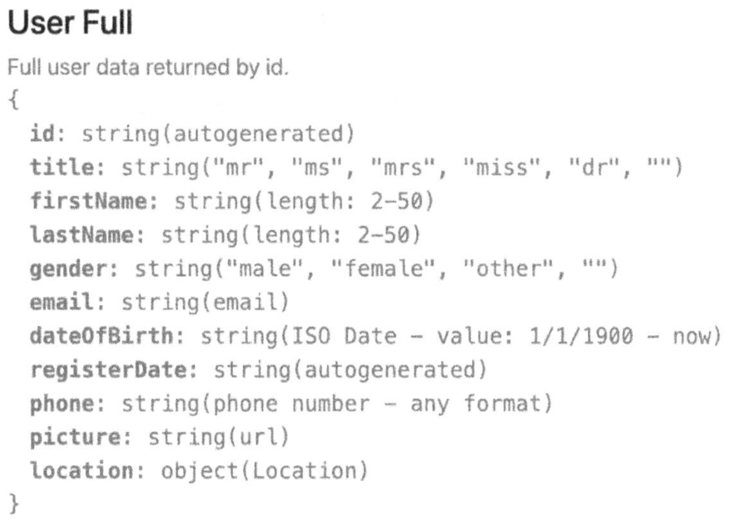

User Full

Full user data returned by id.

```
{
  id: string(autogenerated)
  title: string("mr", "ms", "mrs", "miss", "dr", "")
  firstName: string(length: 2–50)
  lastName: string(length: 2–50)
  gender: string("male", "female", "other", "")
  email: string(email)
  dateOfBirth: string(ISO Date – value: 1/1/1900 – now)
  registerDate: string(autogenerated)
  phone: string(phone number – any format)
  picture: string(url)
  location: object(Location)
}
```

Figure 10.4: Model information for User Full

Model information such as that shown in *Figure 10.4* can help us develop a well-structured code base and make the correct choices ahead of time.

Now that we have a good understanding of the object that the API will return, it is time to build our component. Let's start by defining the HTML section:

```
<template>
<section class="userView">
  <h2>User information</h2>
    <template v-for="key in valuesToDisplay">
     <label v-if="user[key]" >
        {{ key }}
        <input
          type="text"
          disabled
```

```
        :value="user[key]"
      />
    </label>
  </template>
</section>
</template>
```

The HTML markup makes use of the `v-for` directive to loop through a preselected list of properties that we want to display from the user object received from the API. We are then using `v-if` to add some further validation to ensure that our application will not break in case the API changes its returned object. The HTML will generate a simple design that will render a pair of labels/input for each property, but further development could be made to improve the UI.

Next, we are going to create the logic required to fetch and assign our user information. As previously mentioned, we are going to hardcode an ID of 657a3106698992f50c0a5885, which will later be changed to be dynamic:

```
<script setup>
import { reactive } from 'vue';
const user = reactive({});
const valuesToDisplay = [
  "title",
  "firstName",
  "lastName",
  "email",
  "picture",
  "gender"
];
const fetchUser = (userId) => {
  const url = `https://dummyapi.io/data/v1/user/${userId}`;
  fetch(url, {
    "headers": {
      "app-id": "1234567890"
    }
  })
    .then( response => response.json())
    .then( result => {
      Object.assign(user, result.data);
    })
}
fetchUser("60d0fe4f5311236168a109ca");
</script>
```

The API logic is very similar to other requests that we have already made in this book. The only differences are as follows:

- The `url` variable is not a static value but dynamic as it makes use of `userId`.
- Since we are switching the full value of `user`, we had to use `Object.assign`. This is required when changing the entire value of a `reactive` property.
- The value of `valuesToDisplay` has been set to be a simple array and it is not using either `ref` or `reactive`. This has been done on purpose as this array is not expected to be modified and it is therefore not required to be reactive. At the start of your Vue.js development, it is probably best to always define a variable as `ref`/`reactive` as something that you expect to be static could easily turn into dynamic.

The user profile component is now complete, but it is still not accessible from the UI as there is no component or view that loads it. We can fix this in the next section where we will introduce dynamic routes.

Creating a user profile route

The creation of a dynamic route follows the same flow as a normal route. All routes are defined by adding an entry into the `route` array within the route configuration file in `router/index.js`:

```
...
import PrivacyView from '../views/PrivacyView.vue'
import UserView from '../views/UserView.vue'
...,
{
  path: '/privacy',
  name: 'privacy',
  component: PrivacyView
},
{
  path: "/user/:userId",
  name: "user",
  component: UserView
}]
```

Just like before, we added three different properties to our `route` object: name (which will provide us with a friendly name to associate with the route), component (which will show us which component to load – in our case, our newly created `UserView` component), and finally, `path`.

You may have noticed that the `path` value has a different syntax compared to the previous routes and this is what makes this path a dynamic route. In fact, in previous cases, the URL created by the route was unique and static, while in this case, we have added a dynamic part to the URL called **param** (short for **parameter**), and this is defined by the colon (:) preceding the word `userId`. So, this path does not translate into `website.com/user/:userId`, but it defines a URL that expects a value to replace `:userId`. In our example, the path would probably look like `website.com/user/1234` or `website.com/user/my-user-id`.

As we mentioned before, the value that replaced the parameter is then going to be available within the Vue component and can be used to render a unique page. In our case, the ID provided will be used to load the information of a specific user, so accessing URLs with different IDs will render pages with different user information.

> **You can have multiple parameters in a route**
>
> Did you know you can have multiple dynamic parameters in a single route? Each is going to be defined using the same syntax. For example, you could have a URL defined as `/account/:accountId/user/:userId`. The component that will load this route will have access to two dynamic parameters, one containing the account ID and the other the user ID.

Adding navigation using the route name

Now that the user route is defined, it is time to make some changes to our design to ensure that the users can navigate to it.

Even if we could easily achieve this task by using the navigation we learned in the previous section, we are going to take this opportunity to introduce a new syntax that navigates using the route name.

When defining a new route, we provided both a `path` and `name` value for each route item. Navigating using a path is simple for a basic URL, but as things get complicated, it may be cleaner to use the route name to define the navigation code.

Providing a name for every route is a best practice highly adopted in routing systems, as it allows for more readable and maintainable code.

Let's open `SocialPost.vue` and add the navigation logic to the user avatar. To accomplish this task, we are first going to add a `click` event to the `<img>` element:

```
<img
  class="avatar"
  :src="avatarSrc"
  @click="navigateToUser"
/>
```

We attached a method called `navigateToUser` to the `click` event. This method does not have any argument as we will be able to access everything we need directly from the `component` scope.

Next, we are going to create the `navigateToUser` method in the `<script>` block of our component:

```
<script setup >
import { onMounted, ref, computed } from 'vue';
import SocialPostComments from './SocialPostComments.vue';
import IconHeart from '../icons/IconHeart.vue';
import IconDelete from '../icons/IconDelete.vue';
import TheButton from '../atoms/TheButton.vue';
import { useRouter } from 'vue-router';

const showComments = ref(false);
  const onShowCommentClick = () => {
  console.log("Showing comments");
  showComments.value = !showComments.value;
}
const props = defineProps({
  username: String,
  id: String,
  avatarSrc: String,
  post: String,
  likes: Number
});
const router = useRouter();
const navigateToUser = () => {
  router.push({
    name: "user",
    params: {
      userId: props.id
    }
  });
}
```

To add our programmatic navigation, we first imported the `useRoute` method, then we initiated the route with `const router = useRouter();`, and finally, we defined our method.

Navigating using the name utilizes the same method used with the path navigation, `router.push`. The difference lies in the value passed to it. While, in previous examples, we just passed a single string, we are now passing an object.

This object includes the route name and a list of parameters, which, in our case, is equivalent to `userId`:

```
{
  name: "user",
  params: {
```

```
        userId: props.id
    }
}
```

Both the name and the parameters need to match the information defined in the `route` array. This information is case-sensitive, so take extra care when applying it to your navigation object.

Running the application and clicking on the user avatars will redirect to a URL such as `http://localhost:5173/user/60d21b4667d0d8992e610c85`.

Before we can celebrate our work as completed, we need to do one more step, which will require us to read the value from the route and use it to load the correct user. The ID used to load the user from the API has been hardcoded, and clicking on different avatars will result in the same user being displayed.

Reading route parameters in a route component

When a task requires you to create a dynamic route, chances are that you will need one or multiple dynamic parameters to render the page correctly.

If you are creating a blog page, the dynamic parameters will be the blog ID; if you are loading a book library, the parameters could be the category or the author to filter by. No matter the reason, using a dynamic URL will require you to read and use that value within your component.

Reading route parameters can be achieved using the `params` object available within the `route` package. Let's go back to `UserView.vue` and modify the code to load `userId` dynamically from the route:

```
<script setup>
import { reactive } from 'vue';
import { useRoute } from 'vue-router';

const user = reactive({});
const fetchUser = (userId) => {
  const url = `https://dummyapi.io/data/v1/user/${userId}`;
  fetch(url, {
    "headers": {
      "app-id": "657a3106698992f50c0a5885"
    }
  })
    .then( response => response.json())
    .then( result => {
      Object.assign(user, result.data);
    })
}
const route = useRoute();
```

```
    fetchUser(route.params.userId);
  </script>
```

To load our parameters from the route, we made three changes to our view. First, we imported `useRoute` from the vue-router package. Then, we created an instance of the route using `const route = useRoute()`, and finally, we removed the hardcoded `userId` and replaced it with the route parameters using `route.params.userId`.

> **UseRouter versus UseRoute**
>
> You may have noticed that, in previous components, we used `useRouter`, while in the current one, we used `useRoute`. These names are very similar, and their differences can easily be missed. `useRouter` is used when we need to have access to the router object – for example, to add routes, push navigation, and trigger before and after route actions – while `useRoute` is used to gain information on the current route, such as the parameters, path, or URL query information.

At this stage, the dynamic route will be fully functional, and clicking a user avatar will load information for that specific user.

Figure 10.5: User view page displaying user information

In this section, we have learned about all the different aspects of dynamic routes. We have learned why they are needed and the problem that they try to solve, and we introduced their syntax by adding a dynamic route for the users within our `route` array. While doing so, we took the opportunity to learn about programmatic navigation using the `name` and `params` values of a route, and finally, we learned how to access route information, such as `params`, using the route object offered by the vue-router package to make our user view dynamic.

While completing the section, we have also gone over previously learned topics such as components, directives, events, and so much more.

In the next section, we will go one step deeper and learn about nested routes. This will follow a similar flow to the current section and hopefully feel familiar to you.

Learning about nested routes

SPAs are very powerful but the addition of a well-structured router, such as vue-router, can help take SPAs to another level. The strength of a SPA relies on its ability to swap components on the fly without the need to refresh the page, but what if I told you that vue-router can go even deeper than a page layout? This is where nested routes come in handy.

Nested routes provide you with the ability to define routes within routes, multiple levels deep, to create a very complex layout. The concept of nested routes may sound complicated, but their use makes the application easier to develop and they are recommended for most applications.

When we created the main routes, we said that the page would be swapped entirely when on navigation; with nested routes, the concept is the same, but instead of swapping the full page, we just swap an internal part of the page.

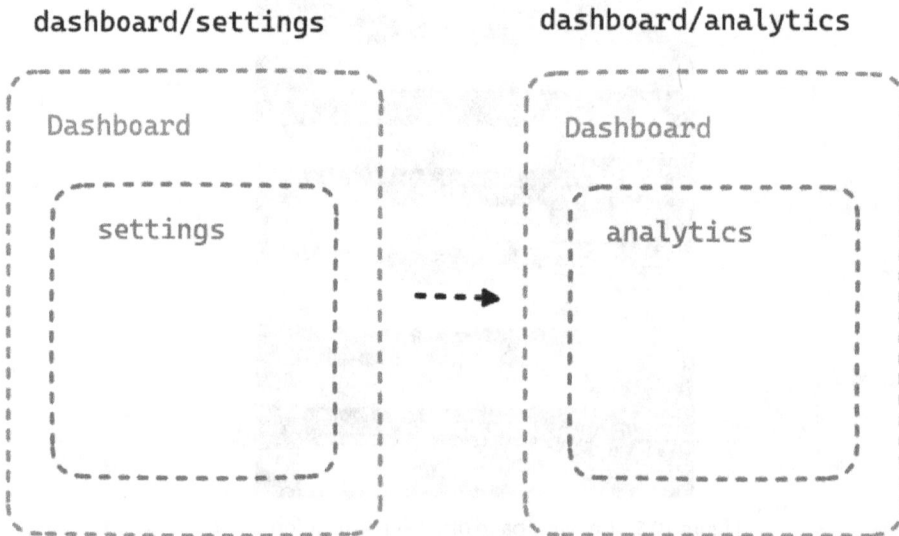

Figure 10.6: Example of nested routes

To better understand how nested routes work, let's discuss the little example shown in *Figure 10.6*. Imagine you have created a dashboard. While developing different parts of the dashboard, such as a settings view and an analytics view, you realize that there is a part of the layout that is shared. This may be the navbar, the sidebar, and even some features such as *print screen* or chat.

To avoid duplicating these functionalities, we can use nested routes. In our example, we would have a parent route called `dashboard`, which will have a set of children – in our example, these will be a `settings` nested route and an `analytics` nested route.

The main `dashboard` route is going to include all the reusable components we mentioned previously, plus the addition of a `<RouterView>` component. This component, just as we did in the `app.vue` file at the start of this chapter, will be used by the router to render the appropriate route.

After this theoretical introduction of nested routes, it is time to apply this knowledge to our Companion App and learn how to use this new feature of vue-router.

Applying nested routes to the user

In this section, we are going back to the user page to enable nested routes by adding the ability to toggle between the user profile and the user posts view.

The example we are going to develop is a very good use case and something that you may encounter in real-life development.

The idea of nested routes is that the views have something in common. This may either be just the layout or some specific route parameters. A very common example of nested routes is an application that has tabs or offers a multi-step form.

Completing this task takes a couple of steps. First, we will create a file that is going to load a list of posts for the user. Next, we will rename the `userView` file to align with the new route, and finally, we will modify the `route` array to define the new nested view.

For the first step, I am going to provide you with just some guidance on how to complete the task, because I want to give you a chance to try and figure it out by yourself. If you get stuck, you can check the `CH10-end` branch, which includes the completed file.

The requirements are to create a file that will load the user posts. This file is going to be called `userPostsView.vue` and will be located within the `views` folder.

The logic of this file is going to be very similar to the `SocialPost.vue` file. In fact, the responses returned by both endpoints are the same, so this allows us to reuse most of the logic and the UI. For the scope of this book, we can copy the content of the `SocialPost.vue` file into our newly created file, but in a real example, we would have refactored the component so that we could have reused it.

The only modification the file needs is to load `userId` from the `route` object, just like we did in `userView.vue`, and then change the URL used to load the information to be `{baseUrl}/user/${userId}/post?limit=10`.

> **Testing and developing a new component**
>
> You may be wondering, "*How can you develop a component if you are not able to test it in the browser?*". It is common practice when developing new components to create dummy routes that can be used to test new components. In this case, our route would need to have access to the user ID from the path, so a simple way to test it out is to swap it in the route array so that the new component is returned when accessing the existing user endpoint.

Next, we are going to rename the userView.vue file to align with the route changes that we are going to make. The new name is going to be userProfileView.vue.

Finally, we are going to learn the syntax required to create a nested route. So, let's open the index. js file within the router folder and see the changes required to split the user endpoint:

```
import PrivacyView from '../views/PrivacyView.vue'
import UserProfileView from '../views/UserProfileView.vue'
import UserPostsView from '../views/UserPostsView.vue'
...
{
  path: "/user/:userId",
  name: "user",
  children: [
    {
      path: "profile",
      name: "user-profile",
      component: UserView
    },
    {
      path: "posts",
      name: "user-posts",
      component: UserPostView
    }
  ],
  component: UserView
}
```

The syntax required for nested routes is quite intuitive and it uses the knowledge that we gained in the previous section.

Nested routes are defined within our existing user route and denoted by the children property. This property accepts an array of routes that is defined with the same structure that we used before with name, path, and component. The only thing that you need to consider here is that when defining nested routes, we are not at the root of the project anymore, so the value assigned to path is added to the existing one. So, in this case, the full path of the userProfile route will be mysite. com/user/:userId/profile.

The user profile and the user posts views are ready to be visited by navigating to `http://localhost:5173/user/:userId/profile` and `http://localhost:5173/user/:userId/posts`, respectively (where `userId` is replaced with an actual user ID).

> **A nested path does not mean a nested layout**
>
> You may have noticed that, with the example in *Figure 10.6*, we mentioned that nested routes can share a layout and be used to update just part of the page such as an inner tab, but in our case, this has not been the case as we are using our nested routes to update the page fully. What we did is not uncommon. In fact, the use of nested routes is not just achieved to share a layout but also to share specific data. In our case, the use of nested routes was used to allow both routes to share the `path` parameters named `userId`. If you want to practice further, you could create a component for the user path and use it as a layout for the user profile and posts route. Just remember to add a `<RouterView>` component in your HTML to define where vue-router should append the routes.

In this section, we learned what nested routes are. We then updated our Companion app to use a nested route to load two different user views. We also created a new view to display user posts. Next, we learned and applied the syntax required to implement nested routes. The final product is the ability for us to navigate to two different pages, one that displays the user profile information and one that displays the user posts.

You may have noticed that we have moved the **user profile view** from the `user` route to one of its children. Because of this, if we access the `user` route (e.g., `http://localhost:5173/user/60d21b4667d0d8992e610c85`), we will be greeted with an empty page. This is not a bug but is expected because we have not actually declared any route that satisfies that specific endpoint. We are going to fix this now by creating an alias.

Reusing routes with alias and redirect

So far, we have learned how to create new routes, but there is another skill required to master vue-router: **alias** and **redirect**.

Both `alias` and `redirect` allow you to reuse existing routes by navigating the user from one route to another and are very useful when you want to create SEO-friendly URLs. The only distinction between the two features is the result that the user sees within the browser.

Using our previous example, we have found ourselves in a situation in which a route is currently unusable because it has no component or view attached to it. This is a common scenario when working with `children` routes and can easily be solved with an alias or redirect.

To solve our empty route problem, we are going to use `redirect` to navigate all users that land on the `user/:userId` path to its `children` path, `user/:userId/profile`.

Creating an alias or redirect is just the same as a simple route, with the only added property of `redirect` or `alias` used to specify what view should be used.

Let's change our user view to include a `redirect` property:

```
path: "/user/:userId",
name: "user",
redirect : { name: "user-profile" },
children: [
```

With the simple addition of the highlighted code, the user is now going to be redirected from one path to another. If you try to access the path `http://localhost:5173/user/60d21bf967d0d8992e610e9b`, you will see the URL change to `http://localhost:5173/user/60d21bf967d0d8992e610e9b/profile` immediately.

The value of `redirect` could either be an object including a name, as in our example, a simple path, or even be a complex object with `name`, `params`, and much more.

It is now time to introduce `alias`. Previously, we saw that we can use `redirect` to transport the user from one route to another. But what if you just want to ensure that the users are able to access a given route using multiple URLs? A good example could be the need for us to create a "friendly URL," which is a URL that is simple for a user to type or remember. It is very common for websites, such as e-commerce websites, to define a specific URL when you are navigating through the website, while also offering a friendly URL that is usually what is indexed and used by Google.

Just to give an example, you may want to render the same route if the user accesses `mywebsite.com/p/123` and `mywebsite.com/product/123`. To achieve this, we can use aliases.

When using aliases, we can define one or more URLs that can be matched with the same route. Aliases accept a string or an array of strings that match the different URLs for a given route.

Let's pretend, for example, that we would like our static `/privacy` page to also render if the user accesses `/privacy-policy`. To achieve this, we would write the following rule in the route:

```
{
  path: '/privacy',
  name: 'privacy',
  alias: '/privacy-policy',
  component: PrivacyView,
},
```

After adding the alias in our route declaration, our Companion app will render the Privacy page for both `localhost:5173/privacy` and `localhost:5173/privacy-policy`.

In this small section, we have introduced the methods of `alias` and `redirect`. We have defined the use case that these two methods solve and their differences. Lastly, we have implemented the method in our own route definition to prevent our user from seeing a blank page when accessing the `user` endpoint and provided a simple example to showcase aliases by defining two endpoints for one route.

Summary

In this chapter, we have learned about vue-router, the first external package that is part of the Vue core set of libraries to be covered in the book. The chapter introduced us to the package by covering its configuration and settings such as history mode. We then learned about the usage of routes, how to define them, and how to navigate to them.

After learning the basics of vue-router, we moved on to more advanced topics such as dynamic routes and nested routes.,

Lastly, we learned about `alias` and `redirect` to complete our basic understanding of the router and how to use it to build both simple and complex applications.

In the next chapter, we are going to learn about another core package of the Vue ecosystem: **Pinia**. This package is used to define and share state across our application.

11

Managing Your Application's State with Pinia

Building web applications is no simple task, not only because of the amount of knowledge required to write them but also because of the architectural complexity that a mature application can develop.

When we first started this book, we introduced simple topics such as the ability to replace text with string interpolation or hide an element with the `v-if` directive. These features are at the core of the Vue.js framework and are needed to build an application with this framework.

As we progressed in the book, we started to introduce topics that are not always required at the very start of your project development – and in some cases not needed at all. In the previous chapter, we covered Vue Router, which was the first additional package to join the Vue.js core framework. We are going to continue the trend by introducing another core maintained package that is part of the Vue ecosystem: Pinia.

Pinia is the official state management package for Vue.js. It is the descendant of the previous state management package, which was called Vuex.

> **Why two different names?**
>
> If both packages have been created and maintained by the same open source maintainer, why do they have different names? Pinia was supposed to be called Vuex 5, but during its development, they decided to give it a different name due to the two versions being majorly different from one another.

As I have already mentioned, some features are not always expected in every application, and state management is one of them. In fact, introducing state management on a very small site would probably be a sign of over-architecture.

In this chapter, we are going to explain what state management is and introduce Pinia as the official package of the Vue.js ecosystem. We will then discuss when an application is expected to include a state management system and cover the pros and cons of using one. We will then do some practice by including two stores within our application: one to handle the sidebar and one to handle our posts. While doing so, we will also add a couple of features to our app, such as the ability to toggle the sidebar from the header and the option to add new posts.

The chapter has the following sections:

- When to use state management
- Learning about the structure of a Pinia store
- Centralized sidebar state management with Pinia

By the end of the chapter, you should be familiar with the notion of state management and be able to define and use stores within your future applications.

Technical requirements

In this chapter, the branch we will use is called CH11. To pull this branch, run the following command or use your GUI of choice to support you in this operation:

```
git switch CH11
```

The code files for the chapter can be found at https://github.com/PacktPublishing/Vue.js-3-for-Beginners.

When to use state management

The first and most important part of this chapter is learning when it is appropriate to use Pinia in your application and when it is not.

All extra packages and features that are added to an application come with an extra cost. This cost is in the time that it takes to learn these new skills, the extra time that a new feature may take to build, the extra complexity that the overall architecture may add to the project, and finally the extra size that another package adds to your JavaScript bundle.

Adding a state manager to your application falls into this category of possible features that may not always be needed. Luckily for us, adding and utilizing Pinia is simple and does not add as much overhead to the project as other counterparts such as React Redux.

The rule of thumb is that state management should only be added to a project if the project is complex enough and includes many layers of components, and if passing values across the application is complicated.

A good use case for Pinia is an application that is large, with properties being passed between many layers. Another use case would be a SPA that has very complex data that needs to be shared and manipulated by multiple parts of the application.

The main problem that state management solves is **prop drilling**. Therefore, the more complex and deep the application structure is, the better suited Pinia is for the project:

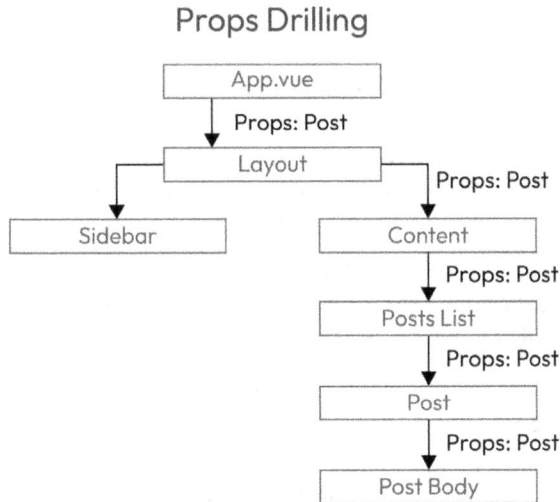

Figure 11.1: Properties are passed down to multiple layers of components

> **What is prop drilling?**
>
> The process of passing data from parents to children, especially multiple layers deep, is also referred to as prop drilling. It is a commonly used term when talking about state management and will also be used in the rest of this chapter.

Let's analyze a few project examples and see whether we would need Pinia's support to handle state management:

- **Brochure site**: A simple site with a couple of static pages, perfect for local businesses. The site may have a contact form. There is no real data that needs to be managed.

  Store not needed

- **A personal blog**: This site is a little more complex, with dynamic pages that render the blog pages as we learned about in the chapter on Vue Router. The data is passed to the pages, but it is not modified or needed in multiple parts of the application.

  Store not needed

- **An e-commerce site**: A commerce site used to sell products. This site will mostly be dynamic, with lots of interactivity offered throughout. Data needs to be passed and modified by many layers of the application, such as from the cart to the checkout.

  Store needed

- **A social media site**: A site that is used to create and share posts. The site will also have pages filtered by user, tags, category, and so on. The same data can be reused in many parts of the application, and using a store can ensure the data is only fetched once and then reused.

  Store needed

Adding a state manager is not a must for all sites, but a requirement that is driven by the specific use case and needs of your SPA. It is true that as you progress in your career and get familiar with the tools and technology that the framework has to offer, you will find yourself over-architecting your application more often than not. However, at the very start of your career, it is important to stay lean and only choose the tools that you really need for your SPA.

Before moving on to the next section, we should say a few more words on the real benefits of using state management. We are going to do so by comparing two applications that have the same component structure but handle data differently. One uses prop drilling, while the other uses state management.

We will start with the example shown in *Figure 11.1*. That example is similar to our Companion App, and it shows the way the `post` property flows from the `App.vue` component all the way down to the last component to render the post on the screen. At this stage, the application is still quite simple. Even if there is some multi-layer prop drilling happening, the complexity is still acceptable.

We are now going to add the possibility for a user to edit a post. Since properties can simply be modified in the component in which they are defined, we will have to emit an "edit" event all the way up the component tree:

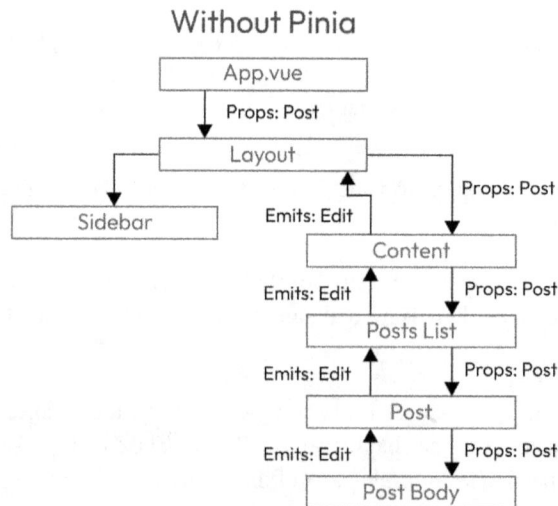

Figure 11.2: Props are passed down and events are passed up the component tree

The situation here is starting to get more complicated. On top of prop drilling, we also have **event bubbling**. The worst part of this scenario is the fact that when a single post changes, the whole component tree will have to be re-rendered because the components are all dependent on the `Post` property.

We are going to include a Pinia store and see what changes it brings to the table. Pinia is going to take ownership of the `post` property and provide it to the component that needs to read or modify it:

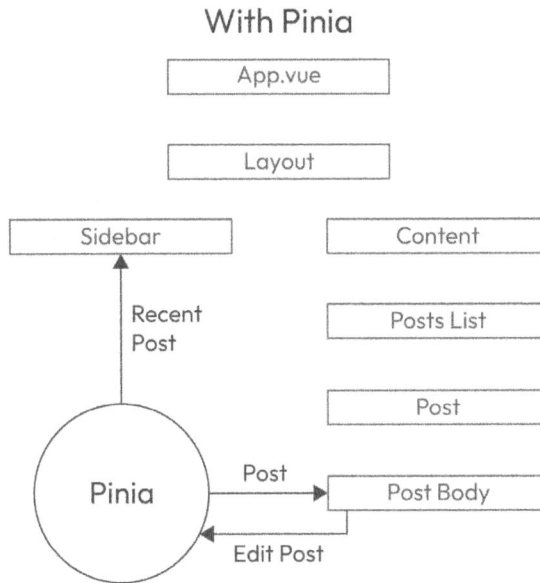

Figure 11.3: A Pinia store managing the data

Adding a store has removed a lot of complexity in our application. When we edit posts, the only component that we re-render now will be **Post Body**. Furthermore, the store easily allows us to pass specific data down, such as **Recent post** in the sidebar. Doing so will also ensure that the sidebar will not be re-rendered unless the post that changed is included in it (this was also possible using props, but was not usually practiced).

In this short section, we defined when state management is needed and when it should be omitted from your application. Then, we defined when a store is needed in your application by describing a few example applications. Finally, we introduced the notion of state management and covered the benefits that it brings to your application.

In the next section, we are going to learn about the structure of Pinia and how we can use it within our application.

Learning about the structure of a Pinia store

In this chapter, we are going to cover what makes up a Pinia store and how it can be used to support you in managing the data of your application. Pinia is built on the notion of multiple stores. Each individual store is going to manage a specific set of data or company logic that is not bound to a specific component. Your application could have a store for posts, a store for comments, and even a store to manage the state of the sidebar.

Stores can talk with each other, but what matters the most is that you should be able to easily define what makes a single store. The stores' data should be split well.

Each store is divided into three different sections: **state**, **getters**, and **actions**.

These three sets of options that are available within a Pinia store can actually compare to existing features that we have learned about regarding Vue Single-File Components.

The **state** object defined in Pinia is comparable to Ref or Reactive used as private component data. The **getters** are comparable to computed properties, as they are used to create a modified version of the existing **state**. Lastly, we have **actions** that are like methods and are used to perform side effect on the store.

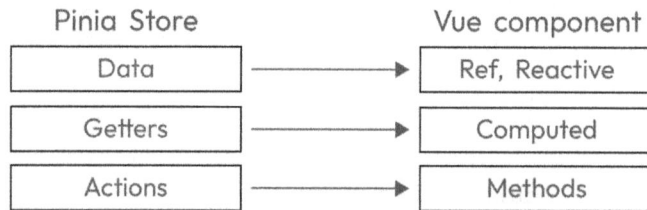

Pinia Store   Vue component

Data	→	Ref, Reactive
Getters	→	Computed
Actions	→	Methods

Figure 11.4: Comparison between Pinia's options and a Vue component

When we first initialized the application, we opted for the `CLI` command to create a store for us. For this reason, the Companion App includes a very simple store example here: `src/stores/counter.js`. Let's see what it includes to learn more about the actual structure of a Pinia store:

src/stores/sidebar.js

```
import { defineStore } from 'pinia'
export const useCounterStore = defineStore('counter', {
  state: () => ({ count: 0, name: 'Eduardo' }),
  getters: {
    doubleCount: (state) => state.count * 2,
  },
  actions: {
    increment() {
      this.count++
    },
  },
})
```

The first part of a store is its declaration. A store is declared using the `defineStore` method available within the `pinia` package. When creating a store, it is common for the exported method to follow the format of **use + store name + Store**. For example, for the store included in the `counter` repository, we can expect the exported method to be named `useCounterStore`.

Next, we have the `state` object. This is declared as a function that returns an object. The syntax may look familiar because it is the same as the syntax that we introduced in *Chapter 2* when we spoke about Vue components being declared using Options API. The `state` object includes the values of the store at their initial state.

Next, we have getters, which are the equivalent of the computed properties available within a Vue component. Getters are used to create a derivative value using `state` or other getters. For example, in a post store, we may have getters for `visiblePost` that just return the posts that have a `visible` flag. Getters receive the state, as the first argument has shown in the previous code snippets within the getters doubleCount.

Last, we have actions. These are equivalent to methods and are used to trigger a side effect that can be used to modify one or more store entries. In our example, the action is called `increment`, and it is used to increase the value of the `count` state. Actions are asynchronous and can include external side effects such as calling an API or calling other store actions.

Now that we have learned how a store is structured, it is time to learn how to use a store within a component. To use a store within a component, we need to initialize it using the exported method generated using `definedStore`. In the instance of the counter store our initialization method would be `useCounterStore` that was defined in a previous code block:

```
const counter = useCounterStore();
```

Then use it to access the state entries directly:

```
counter.count
counter.name
```

The same is then applied for the getters and actions:

```
counter.doubleCount
counter.increment()
```

In this section, we introduced the basic structure of a Pinia store. We learned how to declare it using `defineStore`, then explained the three different parts of a store: state, getters, and actions. Finally, we learned how to access a store from the component by using the counter store to learn the syntax required to access its state, getters, and actions.

In the next section, we will apply what we have learned so far by creating a few stores.

Centralized sidebar state management with Pinia

In the previous section, we introduced the basic structure and syntax of Pinia. To better learn and understand state management, we are going to modify our Companion App by refactoring some of the existing data into its own store. We are going to implement two different stores. The first is going to be a very simple store that will manage the state of the sidebar, while the second is going to handle posts.

The store that handles the sidebar state is going to be quite small. It will be perfect for us to understand the basic syntax and usage of the store, while the one that handles posts is going to be a little more complex.

State management should not be used for all data, and its addition should be accompanied by a good reason that supports its use. So, is it right to add it on the sidebar?

> **Do your research**
>
> Go back to the code base and try to understand everything you can about the sidebar and how it functions. Exploratory knowledge is very useful in developing your tech skills.

The sidebar currently offers these features:

- It can either be opened or closed
- It can be toggled with the use of a button
- It remembers its state using local storage
- All the logic is included within the same component

Our simple `Sidebar.vue` component makes our sidebar nice and interactive. But from the preceding list, one line should catch our eye: "All the logic is included within the same component."

If you jump back to the previous section, you may notice that we mentioned that the use of a store is usually associated with complex scenarios whereby data is passed between multiple components. However, in our case, all the data is stored in just one file and there is not much logic or computation. So why would we need to include a store? And is it a good idea to do so?

The short answer is no. In a scenario like this, a store is not actually needed. Even if I may handle sidebars with a store in my personal projects, I would not suggest that everyone should do so with all Vue projects.

In its current state, the sidebar is too simple for it to be moved into a Pinia store.

Luckily for us, we have full control of the application, so we can simply add a requirement that would make the use of a store appropriate. In this situation, the new requirement is going to be to *add the ability to toggle the sidebar from the main header*.

Even if this requirement seems unreasonable, it is very common for the sidebar to be controlled by a different element. This scenario could become reality for your next project.

To accomplish this new requirement, we would need to perform some prop drilling and event bubbling for it to work without a store. However, with the use of a simple store, the logic is going to be abstracted from the component and easily accessible by the whole application.

The creation of our first store is going to require two steps. First, we are going to refactor our application by moving the existing methods and data into the store. Second, we are going to update the components to use the newly created store.

As we mentioned before, all of the logic that handles the sidebar switching is currently stored in the `Sidebar.vue` component. Within this component, we can find the following code linked to the sidebar toggling:

- Declaration of the `closed` state:

```
const closed = ref(false);
```

- Method to toggle the sidebar:

```
const toggleSidebar = () => {
  closed.value = !closed.value;
  window.localStorage.setItem("sidebar", closed.value);
}
```

- Life cycle to initialize the sidebar state:

```
onBeforeMount( () => {
  const sidebarState = window.localStorage.getItem("sidebar");
  closed.value = sidebarState === "true";
});
```

Let's go and create our first store.

Creating our first store

We are now going to take the preceding code and move it into a new store named `sidebar.js`. Just like we mentioned before, the `Ref` variable is going to change into `State` and the methods are going to become Pinia actions.

Let's start by creating an empty structure for our store:

src/stores/sidebar.js

```
import { defineStore } from 'pinia'
export const useSidebarStore = defineStore('sidebar', {
  state: () => ({}),
  getters: {},
  actions: {},
})
```

Our empty store includes the import of `defineStore` from the `pinia` package, the initialization of the store using the naming convention that we mentioned before (use + store name + Store) that creates `useSidebarStore`, and lastly three empty options for state, getters, and actions.

At this stage, the store is ready to be filled with information. Let's populate it with the sidebar logic:

src/stores/sidebar.js

```
import { defineStore } from 'pinia'
export const useSidebarStore = defineStore('sidebar', {
  state: () => ({ closed: true }),
  getters: {},
  actions: {
    toggleSidebar() {
      this.closed = !this.closed
      localStorage.setItem('sidebar', this.closed)
    },
    loadSidebarFromLocalStorage() {
      const closed = localStorage.getItem('sidebar')
      this.closed = closed === 'true'
    }
  },
})
```

In the preceding code, we declared a new value in the state object called `closed`. This has been set to `true` initially. We have left the `getters` untouched for now, as it is going to be used later in the section. Next, we declared two actions. One was to toggle the sidebar and the other was to load the sidebar from the local storage using existing code from our previous methods.

If we compare the code within the actions with methods that existed in the component, you will notice that they are very similar. The main difference is in the way in which we can access the state. In fact, when these methods were in the component, we had to use `closed.value` to access the value of the ref, while in Pinia, the value of individual state entities can be accessed using the `this.closed` keyword.

Now that our store is complete, we just need to go back into the sidebar and replace the previous logic with the new store. Replacing the current logic with the store will require three steps. First, we need to load and initialize the store. Second, we need to replace the methods with Pinia actions, and finally, we need to modify the template to use the state from the store and not the ref.

Let's start by removing the previous ref and initialize the store:

```
import IconRightArrow from '../icons/IconRightArrow.vue'
import { useSidebarStore } from '../../stores/sidebar';
const currentTime = ref(new Date().toLocaleTimeString());
const router = useRouter();
const closed = ref(false);
const sidebarStore = useSidebarStore();
```

The store is initialized by importing and calling useSidebarStore. This is the exported method that we declared in the store. It is common to declare a constant called either just store or the name + Store.

> **Did you know?**
>
> Using specific names for the store such as sidebarStore rather than just calling it Store can be very beneficial when trying to search for all the usages of a specific store. Since using state management allows you to use this logic anywhere in the app, it is nice to be able to search for it quickly, so having a consistent naming convention is helpful.

In the next step, we are going to work on the methods. We will remove existing methods and replace them with the store actions:

```
const toggleSidebar = () => {
  closed.value = !closed.value;
  window.localStorage.setItem("sidebar", closed.value);
}
const onUpdateTimeClick = () => {
  currentTime.value = new Date().toLocaleTimeString();
};
const navigateToPrivacy = (event) => {
  event.preventDefault();
  console.log("Run a side effect");
  router.push("privacy");
}

onBeforeMount ( () => {
  const sidebarState = window.localStorage.getItem("sidebar");
  closed.value = sidebarState === "true";
  sidebarStore.loadSidebarFromLocalStorage();
});
```

Just like before, the preceding code includes two steps. It first removes the previous logic and then replaces it with the store implementation. We updated the content of the onBeforeMount life cycle by removing the logic that handles the retrieval of the sidebar from the state and replacing it with the loadSidebarFromLocalStorage action.

You may have noticed that we have not replaced the toggleSidebar method yet. This was not a mistake; in fact, we are going to be able to call the Pinia action directly from <template>.

Let's see which changes are needed in the HTML of our component to complete our refactoring into a Pinia store:

src/components/organisms/sidebar.vue

```
<template>
  <aside :class="{ 'sidebar__closed': sidebarStore.closed}">
    <template v-if="sidebarStore.closed">
      <IconRightArrow class="sidebar__icon" @click="sidebarStore.
toggleSidebar" />
    </template>
    <template v-else>
      <h2>Sidebar</h2>
      <IconLeftArrow class="sidebar__icon" @click="sidebarStore.
toggleSidebar" />
      <TheButton>Create post</TheButton>
```

Updating the HTML is the easiest change of all. In fact, the only requirement here is to prepend all states and actions with the `sidebarStore` store constant. Just like actions, state values can also be accessed directly as shown by `sidebarStore.closed`. This was used to access the `closed` state value.

At this stage, our refactoring of the sidebar is complete. All the logic that used to live within the component has been moved into a new store. The sidebar should work as expected, with the only difference being that its value and logic are stored in a store and not in the component itself.

To complete our task, we need to allow another part of the application to toggle the sidebar. This was the requirement that we added to justify our store creation and to learn about the store in more detail.

Expanding the sidebar into the header

In this section, we are going to work in the header to complete our task by adding the ability to toggle the sidebar visibility from a different part of the application.

We are going to do so by adding a simple button next to the settings icon in the header. Just like before, we are going to import and initialize the store and then use its actions directly. It is important to remember that this new button could have been placed anywhere within the application, since its action is owned and controlled by the store.

Let's go into `TheHeader.vue` and add the store:

src/stores/TheHeader.vue

```
<template>
  <header>
    <TheLogo />
```

```
    <h1>Companion app</h1>
    <span>
      <a href="#">Welcome {{ username }}</a>
      <IconSettings class="icon" />
      <IconFullScreen class="icon" @click="sidebarStore.toggleSidebar"
/>
    </span>
  </header>
</template>
<script setup>
import { ref } from 'vue';
import TheLogo from '../atoms/TheLogo.vue';
import IconSettings from '../icons/IconSettings.vue';
import IconFullScreen from '../icons/IconFullScreen.vue';
import { useSidebarStore } from '../../stores/sidebar';
const username = ref("Zelig880");
const sidebarStore = useSidebarStore();
</script>
```

There is no difference between the code we just wrote in the header and the one that we previously defined in the sidebar. In fact, when using a store, we are able to use it anywhere we want in the application without having to define anything else. All instances of the store will work as one, allowing us to use and modify states from different parts of the application.

At this stage, our Companion App will have an added feature that will allow us to toggle the sidebar either by using the sidebar itself or from the header.

Introducing the notion of getters

Before we move to the next section in this chapter, we should introduce another feature of the store that was mentioned but not yet used in our Companion App: getters.

Getters are comparable to computed properties. They allow you to create variables with the use of the store state. In our case, we are going to introduce a simple getter that will create a friendly open or closed label for our sidebar. Outside of this use case, getters can be used for translation purposes, to filter arrays, or to normalize data.

Let's go back to the sidebar.js file and add our friendlyState getter:

src/stores/sidebar.js

```
state: () => (
  { closed: true }
),
getters: {
```

```
    friendlyState(state) {
      return state.closed ? "closed" : "open";
    }
  },
  actions: {
  ...
```

Creating a getter is very simple. You must declare a method within the getters object and then add the logic to create the value that is going to be returned by the getters. Just like computed properties, this is going to be cached. More importantly, it is not supposed to produce any side effects (e.g., calling an API or logging data).

The main point to raise about getters is that they automatically receive the state object as the first argument. So, to access the closed state, we would write `state.closed`. Just like computed properties, thanks to the Vue reactivity system, if the state value changes, the `friendlyState` value will also update automatically.

Now that our getter is in place, it is time to use it. We are going back to the headers and adding this string to be visible right below the user settings. We can reuse the store we previously imported to access the newly defined getter:

```
<template>
<header>
  <TheLogo />
  <h1>Companion app</h1>
  <span>
    <a href="#">Welcome {{ username }}</a>
    <IconSettings class="icon" />
    <IconFullScreen class="icon" @click="sidebarStore.toggleSidebar"
/>
    <p>Sidebar state: {{ sidebarStore.friendlyState }}</p>
  </span>
</header>
</template>
<script setup>
import { ref } from 'vue';
import TheLogo from '../atoms/TheLogo.vue';
import IconSettings from '../icons/IconSettings.vue';
import IconFullScreen from '../icons/IconFullScreen.vue';
import { useSidebarStore } from '../../stores/sidebar';
const username = ref("Zelig880");
const sidebarStore = useSidebarStore();
</script>
```

You may have noticed from the previous code snippets that we did not require any extra initialization or code and that we were able to use the existing store to print the `friendlyState` getters.

The header should now display our string:

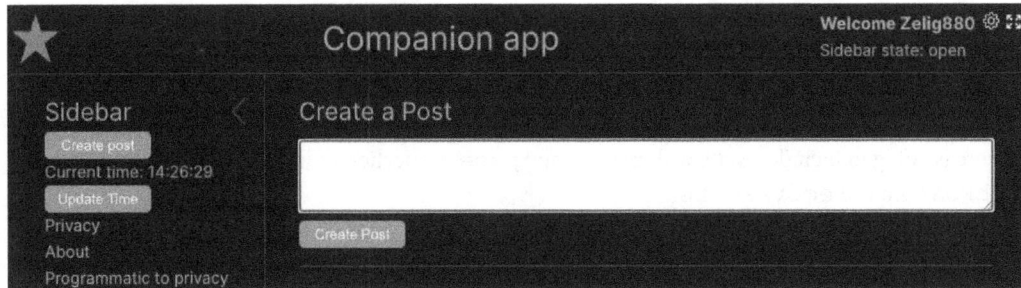

Figure 11.5: Companion App header with the sidebar state

This was a simple example that helped us learn how to refactor existing code, as well as how to define a store with states, getters, and actions. Lastly, it helped us learn how to use the store from one or more components within the app.

We are going to continue our journey to learning about Pinia by introducing another store in our Companion App. In the next section, we are going to create a store that will handle our posts. This is going to be a little bit more complex than the previous one and will allow us to introduce a few more features offered by the state manager.

Creating a post store with Pinia

Adding state management in an application is a never-ending task, as it continues to evolve as the application grows. What we have done so far in our application – refactoring the application by moving logic out of the component and into the store – is a common practice. As we mentioned before, moving the sidebar logic into a store was a bit too much and not expected in a real app, because the logic was small enough to live within the component (even with the prop drilling).

In this section, the situation is different. We are going to refactor a vital part of the application into the store: the posts. Handling the data fetching and management of the posts is a vital part of the application and will probably grow in complexity as your application grows. Because of these points, moving the posts into the store is going to improve the overall application structure.

Just like before, we are going to refactor the current code by first analyzing the code base to find all the code related to the post. We are then going to create a new store and move the code there. Finally, we are going to update the component to use the store.

Since this is the second time we are going through this exercise, I am going to jump past some steps and move directly into the creation of the store. Before you jump to the next section, I suggest you search for all the methods that are related to the post and compare them with the ones that we are going to write in our store. This exercise is going to be extremely valuable as it will provide you with insight into your current understanding of the application.

Our new store is going to be called `posts.vue`. It will be saved in the `src/stores` folder just like our previous store.

This store is going to include a state with `posts` and `page` properties, as well as two different actions: `fetchPosts` and `removePosts`:

```
import { defineStore } from 'pinia'
export const usePostsStore = defineStore('posts', {
  state: () => (
    { posts: [], page: 0 }
  ),
  actions: {
    fetchPosts(newPage = false) {
      if(newPage) {
        this.page++;
      }
      const baseUrl = "https://dummyapi.io/data/v1";
      fetch(`${baseUrl}/post?limit=5&page=${this.page}`, {
        "headers": {
          "app-id": "1234567890"
        }
      })
        .then( response => response.json())
        .then( result => {
          this.posts.push(...result.data);
        })
    },
    removePost(postIndex) {
      this.posts.splice(postIndex, 1);
    }
  },
})
```

We initialized the store by utilizing the naming convention that we introduced before. This produced a named export called `usePostsStore`. Then we declared our state with an empty array for the `posts` variable and a value of 0 for with `{ posts: [], page: 0 }`. Next, we copied our methods and turned them into Pinia's action. The main difference between the previous methods that lived within the component and the copy that we have added to our store is the way in which we access variables such as `page` and `posts`. In a Pinia store, the state values can be accessed using the **this** keyword. So, if a value was previously accessed using `page.value`, we would change this to `this.page`.

This is the only change that we are going to make for the methods, as the rest of the logic stays the same and will not require any modification.

Compared to the previous section, we have yet to introduce anything new, and the refactoring of the post store has followed a very similar flow as the sidebar store implementation. Refactoring logic into a store is usually going to be quite a simple exercise whereby we can lift and shift most of the logic like in our two examples.

Now that the store is set, we will move our focus to the component to ensure it uses the store state and actions. While doing so, we are going to learn how to destructure a Pinia store to increase the readability of our component.

Destructing a store directly using syntax such as `const { test, test2 } = useTestStore()` is not possible, as it would break the reactivity of this state.

Breaking the reactivity would mean that if the value changes within the store, the change will not propagate to the component anymore. To fix this limitation, Pinia exposes a `storeToRefs` method that will allow us to safely destructure the Pinia store.

> **Store or ref is a personal preference**
>
> Using the store directly like we did in our previous example or destructuring values to turn them into Refs is a completely personal preference. There is no right or wrong choice.

Using the store directly will clearly define what data is coming from the store, as it will prefix all data by the store name such as `myStore.firstName`. On the other hand, using Refs will generate a much cleaner component, as the state will not require the prefix, and accessing a state will just require the ref's name such as `firstName`.

Let's complete our store migration by changing the component to use the store and doing so using the `storeToRefs` method. Due to the file including many changes, we are going to break it down into multiple stages:

1. Initialize the store:

    ```
    import { usePostsStore } from '../../stores/posts';
    import { storeToRefs } from 'pinia'
    const postsStore = usePostsStore();
    ```

2. Replace the `private` state with the `store` state:

    ```
    const { posts } = storeToRefs(postsStore);
    ```

3. Replace methods with a store action:

```
const { fetchPosts, removePost } = postsStore;
```

4. Update the HTML.

Due to the fact that we have chosen to turn the store state into Refs, no changes are required within the HTML, as the names of the old Refs and the new Refs now match. The only change that we need to make is in the watch method. In fact, because we turned the posts array from a reactive to a ref, we now need to append the .value for it to function properly:

```
watch(
  posts.value,
  (newValue) => {
    if( newValue.length <= 3 ) {
      fetchPosts(true);
    }
  }
)
```

Before we move on, I want to bring your attention to step 3. In fact, if you were careful, you may have noticed that we extracted the actions of the store directly without using storeToRefs. This is possible because actions are stateless and do not have any reactivity.

The complete file will look like this:

```
<template>
  <SocialPost
    v-for="(post, index) in posts"
    :username="post.owner.firstName"
    :id="post.id"
    :avatarSrc="post.image"
    :post="post.text"
    :likes="post.likes"
    :key="post.id"
    @delete="removePost(index)"
  ></SocialPost>
</template>
<script setup>
import { watch } from 'vue';
import SocialPost from '../molecules/SocialPost.vue'
import { usePostsStore } from '../../stores/posts';
import { storeToRefs } from 'pinia'
const postsStore = usePostsStore();
const { posts } = storeToRefs(postsStore);
const { fetchPosts, removePost } = postsStore;
```

```
watch(
  posts.value,
  (newValue) => {
    if ( newValue.length <= 3 ) {
      fetchPosts(true);
    }
  }
)
fetchPosts();
</script>
```

Implementing an Add Post action

A few chapters ago, we introduced a component aimed at adding new posts to our state. This component was never fully implemented, as it is missing the main logic required to create a post. The reason why the component was left in this state was that adding the functionality without a store in place would have required a lot of prop drilling and event bubbling.

Thanks to the post store, this is not the case anymore. In fact, we are going to be able to generate the logic required to add a new post using store actions.

The store is the sole owner of the posts, so we do not have to worry about where the posts are used. We can simply create an action that adds a post knowing that the store will handle the propagation and handling of the state within the rest of the application.

First, we are going to add an action in our posts.js file:

src/stores/post.js

```
addPost(postText) {
  const post = generatePostStructure(postText);
  this.posts.unshift(post);
}
```

The addPost action is going to add a post by adding it at the start of the posts list. For the scope of our Companion App, adding a new post will just set the main body of the post, because other information such as ID and user information are going to be hardcoded and provided by a function called generatePostStructure..

Next, we are going to initialize the store and attach this action to the createPostHandler:

```
<script setup>
import TheButton from '../atoms/TheButton.vue';
import { usePostsStore } from '../../stores/posts';
import { onMounted, ref } from 'vue';
```

```
const postsStore = usePostsStore();
const { addPost } = postsStore;
const textareaRef = ref(null);
const createPostForm = ref(null);
const createPostHandler = (event) => {
  event.preventDefault();
  if(createPostForm.value.reportValidity()){
    addPost(textareaRef.value.value);
  };
}
...
```

Using the new action follows the same syntax used before. First, we import the store. Second, we initialize it. Third, we destructure the actions that we want to use, and lastly, we use the action as if it were a simple method.

The preceding example uses the text area ref to get the `textarea` value. This is not the correct use of Vue.js and accessing values like this should be avoided. In fact, in the next chapter, we are going to refactor this file by introducing two-way binding with `v-model`.

> **Create your own store**
>
> Before moving on to the next chapter, you should try and create your own store. You can create a very similar store to the sidebar by creating a store that can handle the visibility of the `create post` component. You can use the button within the sidebar labeled **Create Post** to toggle the `CreatePost.vue` visibility. You can see the full implementation in the `CH11-END` branch.

In this section, we continued learning about the Pinia store by refactoring the post data. We created a new store by defining its state and actions. We then converted the existing code to work within the Pinia store. Next, we introduced the `storeToRefs` method and learned how to destructure state and actions from a store. Finally, we made use of the new store by adding the ability for users to add a new post by simply creating a new action. What you learned here is not a complete list of features offered by Pinia, but a quick introduction to a great state management package. As you practice more, you will then learn about other features such as `$patch` and `$reset`.

Summary

Introducing state management into your application can really help you handle your data with ease. In the two examples that we shared in this chapter, we saw the benefit that state management can add to your application by avoiding prop drilling and event bubbling.

Before we conclude this chapter, I want to share one more benefit of using state management in your application. The refactor that we have accomplished in the previous two sections highlights the fact that using the Pinia store helped us remove lots of logic from the components to a single location within the store file.

This abstraction is not only good for a development experience but can also be used to choose which parts of our application can be unit tested. You may remember from *Chapter 8* that choosing what to test is quite complicated, as there is a very fine line between testing too much and testing too little. I personally use state management to delineate which parts of the application I will unit test. I achieve this by always making sure all stores are thoroughly tested.

In this chapter, we first introduced the notion of state management and talked through the syntax and features that Pinia has to offer. We then started to put what we learned into practice by refactoring the sidebar into its own store. By doing so, we learned how to declare state, getters, and actions and how to use them within our components. Next, we continued our learning by refactoring one more piece of our application: the posts. We created a store and converted the methods into Pinia actions. Finally, we learned how to destructure state and the importance that state management can have in our application architecture.

In the next chapter, we are going to learn how to handle forms in our application by introducing two-way binding with `v-model` and client-side validation with **VeeValidate**.

12

Achieving Client-Side Validation with VeeValidate

A website is not complete until it has a form of some sort. Contact Us, feedback, and comment forms are just a few examples that we may be required to develop and validate in our Vue.js application.

In this chapter, we will explain what makes a semantically correct form, introduce the available fields, and talk through its accessibility needs. We will then update our `CreatePost` component and learn how to handle form fields and manage two-way binding using `v-model`. Next, we will move on to creating a new form called `contactUs`. This will be used to introduce a new package called **VeeValidate**. Finally, we will learn how to develop custom validation rules and how to simplify our validation by using VeeValidate preset rules.

In this chapter, we will cover the following:

- Understanding forms
- Two-way binding with `v-model`
- Controlling your form using VeeValidate
- Defining your form validation with VeeValidate

By the end of the chapter, you should have gained a good understanding of forms and how to handle them in Vue.js. You will be able to define well-structured and accessible forms, handle user input with `v-model`, and define complex forms with validation using VeeValidate.

Technical requirements

In this chapter, the branch is called CH12. To pull this branch, run the following command or use your GUI of choice to support you in this operation:

```
git switch CH12
```

The code files for the chapter can be found at https://github.com/PacktPublishing/Vue.js-3-for-Beginners.

Understanding forms

Whether you are new to development or you already have some experience, it is important to spend a bit of time looking at what makes a good form and how to best define forms.

I started to learn forms over 10 years ago but still, for me, there was something new to learn while I completed my research for this chapter. HTML5 enhancement and accessibility requirements have modified the way we define forms. In this section, we are going to understand what makes a good form, and then, in later sections, we will use this knowledge to define some forms in our Companion App.

In most static sites, such as brochure sites or blogs, forms are the only time in which a user interacts with your site, and as such, they need to offer a great user experience (UX) and be as accessible as possible.

A good form includes three different aspects:

- It is semantically correct
- It is accessible
- It is validated

By the end of this chapter, we will have covered all three aspects, and you will be able to create great forms.

Let's start with the first point and see what makes a good HTML structure. Using the correct HTML is not only going to improve our UX but it is also going to set the foundation for the accessibility work that will be covered later in the section.

Please note that this book is not about basic HTML and JavaScript but it focuses on Vue.js. For this reason, this section is just going to be a very quick introduction to forms; we will not spend too much time going into the details but will cover the main topics so that you can follow the rest of the chapter at ease and ensure that your form is well structured.

Wrapping your form within the <form> element

Let's start with the basics that are often overlooked. All forms need to be wrapped within a <form> element. This is not just something nice to do but it also has some very important meaning behind it.

Using the `<form>` element has three benefits. First, it informs the browser about the presence of a form (important for browser extension). Second, it improves the UX for visually impaired users who use a screen reader (software that reads the page for users) by supporting **Form** mode (a specific mode used by screen readers to complete forms on websites). Next, it helps to handle validation.

You may be asking yourself how the use of the `<form>` element improves UX. Well, have you ever wondered why, sometimes, a form can be submitted by pressing the *Enter* key and sometimes it can't? Well, this is driven by the `<form>` element. When a form is wrapped in a `<form>` element, if the user clicks the *Enter* key while completing it, the form will trigger its `submit` method.

Do not forget labels

The topic of the `<label>` element holds a special place in my heart. Many developers and designers miss the importance of this element and either remove it from the UI or misuse it. There was a trend a few years ago to develop compact form designs in which the placeholder replaced the label. These forms seemed very clear and started to be used everywhere, but they were a big failure from a UX perspective. Even if they look great, they can be hard to use because the placeholder disappears as soon as the user (or the browser) types something in the field and cannot be seen again until the user removes the entry.

Another problem with labels is the fact that they are essential from an accessibility perspective. In fact, without them, a visually impaired user would not be able to know what the field is about and, therefore, would be unable to fill in the form completely.

A label can be used in two different ways. It can either be used as a standalone element, using the ID to link it to an input field, or it can wrap the input field to which it belongs:

```
// Label standalone
<label for="name">Name:</label>
<input type="text" name="name" id="name">
// label wrapping input
<label>Email:
   <input type="email" name="email" id="email">
</label>
```

Both methods are semantically correct, and their use is usually driven by the design that you may need to achieve.

There is more than just type="text"

As I mentioned before, I started using forms a long time ago, and one thing that has progressed over time is the different types of input available nowadays and supported in all major browsers.

This is the list of all available input types:

- `button`
- `checkbox`
- `color`
- `date`
- `datetime-local`
- `email`
- `file`
- `hidden`
- `image`
- `month`
- `number`
- `password`
- `radio`
- `range`
- `reset`
- `search`
- `submit`
- `tel`
- `text`
- `time`
- `url`
- `week`

Browsers offer fields for telephone numbers, dates, and even for email. But what would be the benefit of using them? Why not just use a simple `text` field? Well, the difference may not be in the way the field looks but, rather, in the way the field works.

Different fields may trigger different UIs. For example, a `date` field will open a calendar when clicked, a `number` field might show little toggles, and so on. This improvement in UX is even more noticeable on mobile phones where handsets can provide well-defined native components to support users in completing their forms.

Setting your form to be autocompleted

Who doesn't love a form that is completely autocompleted by the browser? Luckily for us, browsers are working very hard to complete as much of the form as possible out of the box. There are a couple of things that we can do to help browsers.

First and foremost, we need to inform the browser that we would like the form to be autocompleted. This is not done automatically but requires the `<form>` element to receive an `autocomplete` attribute:

```
<form autocomplete >...</form>
```

Next is ensuring that we describe the field with the correct name, as browsers can read the name and assign the correct field. So, if a field is an address, you should give it the proper name of `address` and not use shorthand such as `addr` or something else that the browser cannot understand.

Last but not least is defining the correct value to assign to the autocomplete attribute. Setting `autocomplete` on the main form may not be all that is needed. Sometimes, as in the case of browser extension used to autocomplete your passwords, the browser requires further information, and this can be passed using the `autocomplete` attribute directly on the field. `autocomplete` can accept different values depending on the role of the input. Here are some examples:

- For a username:

  ```
  <input type="email" autocomplete="username" id="email" />
  ```

- For a sign-up form:

  ```
  <input type="password" autocomplete="new-password" id="new-
  password" />
  ```

- For signing in:

  ```
  <input type="password" autocomplete="current-password"
  id="current-password" />
  ```

As shown by the username example, the type and name of the input do not have to match their `autocomplete` value. In fact, in the case of the username, even if the browser sees and uses the field as an email, a password tool would fill the field with the username.

In this section, we have learned how to define a good form. Wrapping the form in `<form>`, setting `autocomplete`, defining the correct labels for each field, and using the correct input type are just a few steps that we can take to ensure that our users have a great experience in filling out our form.

In the next section, we will learn how to define and use forms in Vue.js and how the framework can help us improve the UX further.

Two-way binding with v-model

In the previous section, we learned how our form should be defined, and it is now time to put this learning into practice by enhancing our Companion App with a couple of forms. In this section, we will learn how to use `v-model` to enable our input field to accomplish two-way binding, which is a term used to describe when a field can emit change events and update values simultaneously (hence two-way).

In the previous chapter, we mentioned that the solution used to manage the value in the `CreatePost` component was suboptimal, and it is now time to align it to use the best industry standards.

Let's start by rephrasing what two-way binding really achieves. Let's consider an input field, for example. Until now, we have learned how to set the value of this field using `ref`. The use of `ref` allows us to set the value of a specific input field at load, but this is not very useful when it comes to forms, as we want the value to be set by the visitor's input. In the previous chapter, we learned how to use events to emit change, so in the case of an input field, we could emit a change event that changes the `Ref` variable and, in turn, updates the value of the field.

This would look something like this:

```
<input value="firstName" @input="firstName= $event.target.value" />
```

Even if the preceding code does achieve two-way binding, it does not really look clean, and adding this all over our forms would make the HTML very hard to read.

Luckily for us, the Vue.js framework has a nice shortcut that will make our HTML clean, called `v-model`.

Changing the preceding to use `v-model` would result in the following syntax:

```
<input v-model="firstName" />
```

Behind the scenes, `v-model` and the `event/value` syntax produce the same code, but due to the clean and concise code of `v-model`, it is the suggested option to achieve two-way binding when possible (there may be cases in which we would need to handle events manually to handle specific side effects; in these scenarios, the `event/value` syntax may be preferred).

Now that we have learned about the syntax and how to define the form, let's go back and apply our learning to the `CreatePost` component:

```
<template>
  <form
    v-show="postsStore.inProgress"
    ref="createPostForm"
    @submit.prevent="createPostHandler"
  >
    <h2>Create a Post</h2>
    <label for="post">Enter your post body:</label>
```

```
    <textarea
        rows="4"
        cols="20"
        ref="textareaRef"
        required="true"
        minlength="10"
        v-model="postText"
        name="post"
        id="post"
    ></textarea>
    <TheButton>Create Post</TheButton>
  </form>
</template>
<script setup>
import TheButton from '../atoms/TheButton.vue';
import { usePostsStore } from '../../stores/posts';
import { onMounted, ref } from 'vue';
const postsStore = usePostsStore();
const { addPost } = postsStore;
const textareaRef = ref(null);
const createPostForm = ref(null);
const postText = ref("");
const createPostHandler = (event) => {
  event.preventDefault();
  if(createPostForm.value.reportValidity()){
      addPost(textareaRef.value.value);
  }
  addPost(postText.value);
}
onMounted( () => {
  textareaRef.value.focus();
});
</script>
```

Let's break down all the changes we applied to our form. First, we removed the Ref initialization of createPostForm. This reference was made to hold the value of the HTML form, but this is no longer needed after the refactor. Next, we created a reference called postText that will hold the value of our input field. We then added v-model within our input field, assigning it to the newly created postText.

At the first load, the input field is going to be an empty string (const postText = ref("");), but this value is going to be automatically updated by v-model as the user types into the input field. Behind the scenes, v-model assigns the value on the first load and then automatically updates it every time the input field emits a change event.

While refactoring the form, we also introduced a small enhancement by adding a `.prevent` modifier to the `@submit` event. Doing so automatically calls `event.preventDefault()` when submitting our form.

The last change is in the payload of the `addPost` method. This method is now using `postText.value` instead of using the input reference to get the value.

> **Quick reminder**
>
> Just a quick reminder that using Template Ref to get an input value was just used for teaching purposes and should not be used in real life. Using Template `Ref` to access the HTML object should just be used for methods and actions that cannot be achieved with basic Vue.js features such as directives, computed properties, and events.

At this stage, the form is fully working using Vue.js two-way binding supported by `v-model`. If we try to use the form, this will work as expected and either create a post or show the browser native validation on incorrect input, as shown in *Figure 12.1*.

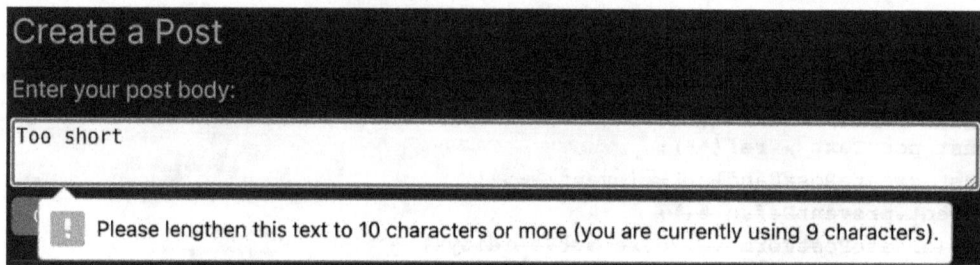

Figure 12.1: Native validation message shown in the Create a Post form

With the knowledge you have gained in this section, you will be able to create forms and handle user input within your application. In many scenarios, the validation offered by the native browser is either not "pretty" enough or does not fulfill the validation requirement of the form.

For this reason, in the next section, we are going to learn how to use an external package called VeeValidate to control the validation of our forms.

Controlling your form using VeeValidate

In the previous section, we learned how to handle user input such as form fields using `v-model`, but in real life, handling forms is a little more complicated than just setting two-way binding.

In fact, a complete form will include complex validation and error handling, just to name a couple. Even if we could achieve this feature manually using Vue.js, we would probably end up investing a lot of time and effort in something that is well handled by an external package. For this reason, in this section, we are going to introduce **VeeValidate**, which is a package aimed at making form development easy.

On the official website (`https://vee-validate.logaretm.com/`), VeeValidate is described as follows:

> *"VeeValidate is the most popular Vue.js form library. It takes care of value tracking, validation, errors, submissions and more."*

Using an external package is not required for every single form you will ever write, but it is useful if you want to create a well-written form, as the package provides you with a consistent way to define and handle logic.

The first step when dealing with a Node package is its installation. To install VeeValidate, we will have to open the terminal at the root of our project and run the following command:

```
npm install vee-validate
```

This command will add the package to our repository for us to import and use.

In the next step, we will create a form and learn how to use this package to make form development easy and consistent. We could have reused the **Create a Post** form, but we will create a new one so that you have both examples to look at in the future within your code base.

Some of the scaffolding for this new form is already in place. We have a new button on the sidebar with the label **Contact Us**, a new page added to the `router` object, and a Vue.js component called `ContactView.vue` that uses the static template we defined in a previous chapter. Last but not least, we also have a basic page scaffolding with a component called `ContactUs` that is going to be our playground for the creation of this new form.

To better understand VeeValidate, we are going to build our form in steps, starting from the basic form creation:

```
<template>
  <form @submit="handleSubmit">
    <label for="email">Email</label>
    <label for="message">Message</label>
    <TheButton>Send</TheButton>
  </form>
</template>
<script setup>
import TheButton from '../atoms/TheButton.vue';
const handleSubmit = ({email, message}) => {
  console.log("email:",email)
  console.log("message:",message);
};
</script>
```

We created a `<form>` element with a `submit` event attached to a method called `handleSubmit`. We then defined two labels, one for `email` and the other for `message`, and finally, a button that will be used to submit the form.

You may have noticed that we do not have any input fields yet. These were omitted on purpose because they are going to use built-in components offered by VeeValidate.

Using VeeValidate will take full control of the form, including its state, meaning that we will not have to define or manage the individual values as we did in the previous example with `v-model`.

The way in which VeeValidate can make this happen is with the use of built-in components that are purposely built to handle and manage the form for us.

In this section, we are going to look at three components: `Form`, `Field`, and `ErrorMessage`.

The `Form` component is going to replace the native `<form>` element available in HTML. `Field` is going to take the place of the `<input>` field, and finally, `ErrorMessage` is going to be used to display a custom message if the field enters an `error` state.

Let's go back to our form and update it to use the new built-in components and then talk through how they got used:

```
<ｆForm @submit="handleSubmit">
  <label for="email">Email</label>
  <Field
    id="email"
    type="email"
    name="email"
  ></Field>
  <label for="message">Message</label>
  <Field
    id="message"
    as="textarea"
    name="message"
  ></Field>
  <TheButton>Send</TheButton>
</ｆForm>
</template>
<script setup>
...
```

Our form is now updated to use VeeValidate. The first change may not be very noticeable, but it includes the change of the `<form>` element to the VeeValidate custom element, `<Form>`. The custom `<Form>` element is used by VeeValidate to handle the form values' state.

Next, we accompanied each label with a `Field` component. This component is imported from the `vee-validate` package.

The `Field` component is very similar to native form elements such as `<input>` and `<radio>` and accepts most of the attributes that you would use in a native input, such as placeholder and name type.

We created two fields. The first field is used for the email and has three attributes: `id`, `type` (which is `email`), and `name`, which is going to be used to connect the field with its error messages.

The second field is associated with `message`; it is similar to the previous field used for the email, with the only difference being an extra attribute called `as`. This property is used to specify which form element the field should be rendered as. When using the `Field` component without using the `as` property, it defaults to an `<input>` element. In fact, for our message, we want the field to be a text area, and we achieved this by using the `as` property and assigning it a value of `textarea`.

At this stage, the form has the HTML and the logic required to function. If we open our application at `http://localhost:5173/contact` and fill in the form, we will be able to see the form being submitted in the console:

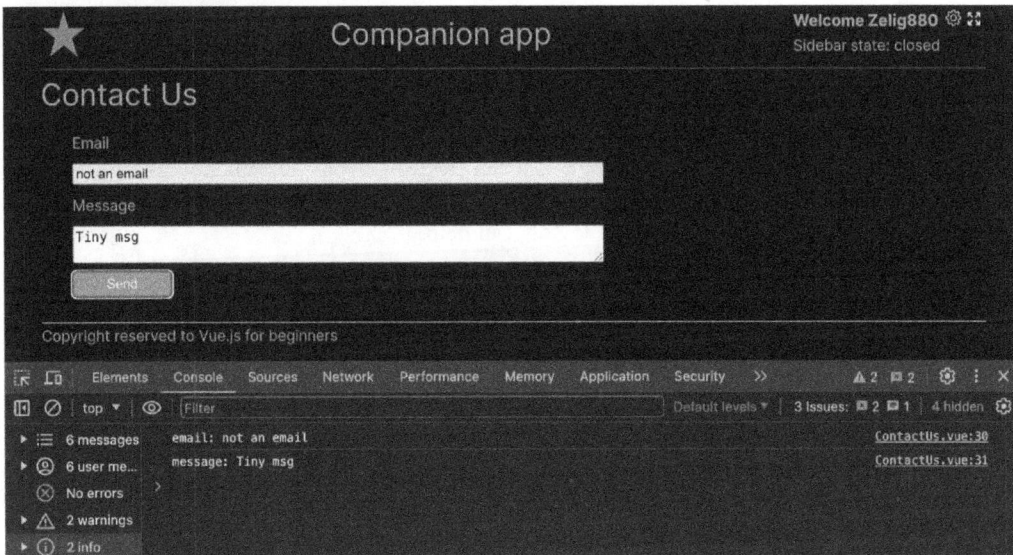

Figure 12.2: Console message triggered by the Contact Us form

Even if we have not finished yet with VeeValidate, you should be able to see the benefits that it brings. You may have realized that we never had to declare any references for the email and message or define any two-way binding with `v-model`. All of this is handled behind the scenes by the package with the `Form` and `Field` custom fields.

In this section, we have added VeeValidate to our form and learned what benefits this will bring to our application. We then refactored our **Contact Us** form to use the newly installed package by introducing the `Field` and `Form` components.

We have just one more modification to apply to our form before we can complete this chapter: validation. The preceding example shows that I was able to submit a form with incorrect values such as a fake email and a very short message. In the next section, we are going to learn how to use form validation with VeeValidate.

Defining your form validation with VeeValidate

A form is never complete until there is some sort of validation. The values of the form are sent to our servers, and we want to make sure that our forms can help the user with immediate feedback if any of the field values are incorrect.

In our case, the form we used in *Figure 12.2* would have failed backend validation as the email was not in the correct format. In this section, we are going to learn how to create custom validation and introduce another supporting package offered by VeeValidate that includes a preset group of validation rules to speed up our development.

> **Frontend validation is not enough**
>
> Remember that frontend validation like the one performed by VeeValidate is just going to improve the UX but it is not secure enough as it can easily be bypassed. When working with forms, you should also define validation on the backend. Using validation schemas that work on both the frontend and backend, such as Zod, can help.

VeeValidate offers the possibility to define its rules using either a declarative approach or by using composition functions (a set of composable functions offered by VeeValidate that can used to create your own form components). For the scope of this book, we are going to use the declarative method.

To validate our input fields, we need to define some validation rules. These rules are going to be methods that run a specific condition on one or more input fields. An example could be a `required` validation that would check whether the value of an input is defined, or a *minimum characters* rule that would check whether the number of characters is equal to or greater than the limit set.

The first rule we are going to define is a simple `required` rule. To define a rule, we use a VeeValidate method called `defineRule`. This method accepts two arguments. The first is the name of the rule and the second is the function that will run to assess the validation. We can add the following code within the `ContactUs.vue` file:

```
<script setup>
import { Field, Form, ErrorMessage, defineRule } from 'vee-validate';
import TheButton from '../atoms/TheButton.vue';
defineRule('required', value => {
  if (!value || !value.length) {
```

```
    return 'This field is required';
  }
  return true;
});
```

Declaring a rule validation is not any different from a method that you would have defined natively in Vue. In the case of our required rule, we are checking whether the value is set and whether it has at least one character by using `!value` and `value.length`, respectively. Then, we are either returning `true` (if the validation passes successfully) or we are returning a string as an error message.

After declaring the rule, all that is left is to use it within a form field. Validation rules can be assigned to the `<Field>` element that we used before. The `Field` component expects a property called `rules`. This property can either accept a string or an object. I prefer the string notation as it is like the native validation in HTML but also because it has a shorter syntax that helps keep the `<template>` section clear and readable.

Applying the `required` rule to our `email` field would produce the following code:

```
<Field
  type="email"
  name="email"
  placeholder="Enter your email"
  rules="required"
></Field>
```

At this stage, our form will not be submitted anymore unless the validation passes successfully. In fact, VeeValidate will just trigger our `submit` method if all fields in the form are valid. The next step required is to display an error message to the user when the form is invalid. This can be achieved with another component offered by VeeValidate, called `ErrorMessage`. The error message itself has already been defined in our validation rule, and VeeValidate is going to take care of the logic required to display and hide the message.

The `ErrorMessage` component accepts a property of name that is used to connect it with a specific input field. So, in our case, it is going to be `name="email"`:

```
<Label for="email">Email</Label>
<Field
  type="email"
  name="email"
  placeholder="Enter your email"
  rules="required"
></Field>
<ErrorMessage name="email" />
```

When trying to submit the form, we will now be presented with an error message, as shown in *Figure 12.3*.

Figure 12.3: Form displaying a validation error

Even if our error is displayed successfully, it does not really stand out as it uses the same color as the rest of the text on the page. This is happening because the `ErrorMessage` component provided by VeeValidate is just a utility component, meaning that it does not actually provide any HTML markup but it is used to provide some additional functionality such as handling the error placement and visibility.

To improve our UI, we can use the `as` property to define an HTML element that will be used to wrap the error message, just like we did with `textarea`, we can use the attribute "as" to define what HTML field the component should be rendered as and add some classes. The code base already has some styles for a class named `error`. So, let's go and modify our `ErrorMessage` component to use this class:

```
<ErrorMessage as="span" name="email" class="error" />
```

After adding the `as` property and `class` attribute, our component is now going to be more prominent on the screen.

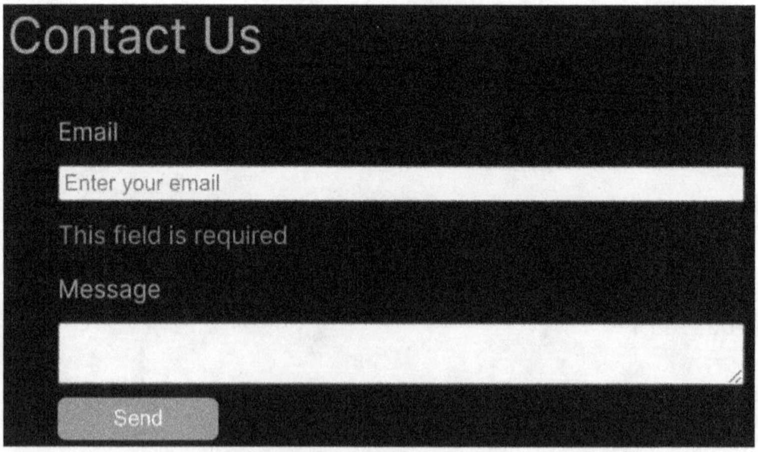

Figure 12.4: Form showing a red error message

The form is starting to take shape. The form fields and labels are set, the submit handler is in place (even if just a dummy one), and the validation rules are defined and working.

In the next and last section of this chapter, we are going to learn about preset validation rules offered by VeeValidate. Declaring a simple rule such as required is simple, but this is not always the case.

I personally love to use preset rules as they help me keep components small and readable while still being able to provide complex validation rules.

Using VeeValidate rules

So far, we have just applied a simple validation to the email field that checks whether it is set, but this would still break as the field is currently going to pass validation with just a single character.

In this section, we are going to validate the email (to ensure it is written in a valid format) and the message text area (to ensure that we receive a message that is at least 100 characters).

Instead of declaring the validation manually by defining its rules, we are going to use the @vee-validate/ rules package and use two of its preset rules to achieve our validation requirements.

VeeValidate offers over 25 rules (https://vee-validate.logaretm.com/v4/guide/ global-validators#vee-validaterules), including simple rules such as required, numeric, and emails, as well as complex ones such as one_of, not_one_of, and regex.

Before learning the syntax required to use these rules, we need to install the package in our application. We can do this by running the following:

```
npm install @vee-validate/rules --save
```

The power of VeeValidate and its validation is that it allows us to use more than one validation for a single field, allowing us to define complex rules, such as the one defined by password fields, without the need to write a single line of code.

To achieve our validation needs, we are going to use the email and min rules:

```
<template>
<Form @submit="handleSubmit">
  <Label for="email">Email</Label>
  <Field
    type="email"
    name="email"
    placeholder="Enter your email"
    rules="required|email"
  ></Field>
  <ErrorMessage as="span" name="email" class="error" />
  <label for="message">Message</label>
```

```
  <Field
    as="textarea"
    name="message"
    rules="required|min:100"
  ></Field>
  <ErrorMessage as="span" name="message" class="error" />
  <TheButton>Send</TheButton>
</Form>
</template>
<script setup>
import { Field, Form, ErrorMessage, defineRule } from 'vee-validate';
import TheButton from '../atoms/TheButton.vue';
import { required, email, min } from '@vee-validate/rules';
defineRule('required', required);
defineRule('email', email);
defineRule('min', min);
...
```

Let's analyze our latest code snippets and see how to use validation rules from VeeValidate.

First, we manually imported the validation from the `@vee-validate/rules` package. We could have imported all rules globally, but I prefer to do it within each component. This not only helps me understand the validation used within the component's scope but also helps keep the built size small by ensuring we import the rules that we use within the code base.

Then, we removed the `required` rule that we defined previously and replaced it with the ones offered by VeeValidate. We then declared two new rules called `email` and `min`. We did so using the `defineRule` method with the imported rules from `@vee-validate/rules`.

Next, we have added our validation within the `<template>` section of our component. The `rules` property can accept multiple entries separated by the character `|`. In the case of `email`, the rules are `"required|email"`, but for `textarea`, they are `"required|min:100"`.

Lastly, we added `ErrorMessage` to our message text area.

You may have noticed that the rules used in the `message` field are a little different from the ones used in `email`. This is because the `min` rule requires an argument that is equal to the number of characters required before the field can be marked as valid. To set a specified parameter for a rule, we add a colon (`:`) followed by one or more values separated by a comma. For example, to define a max rule of `164` characters, we would write `rules="max:164"`, while to define a rule that allows a number between `10` and `20`, we would write `rules="between:10,20"`.

Rules can be defined as an object

As I mentioned before, the `rules` property can be defined as an object. What you choose is personal preference and there is no right or wrong. If we wanted to replace the `message` field rule to be an object, we would write `rules="{ required: true, min: 100}"`.

In this section, we have learned how to validate our form. We first learned how to define and use validation rules by creating a simple `required` validation, and then we introduced the preset validation rules and used them to fully validate our form.

Your turn

Continue to work on the **Contact Us** form by adding more fields with complex rules. This will help you navigate through the documentation and ensure that you have understood the process required to create and handle forms. Lastly, you should enhance the `handleSubmit` method to send a `fetch` request (to a fake endpoint) with the form input.

Summary

In this chapter, we learned how to handle user interaction by introducing two-way binding using `v-model` and form validation and handling using VeeValidate. During the chapter, we redefined what makes a semantically correct form, learned how to use the `v-model` syntax to define two-way binding, and then moved into the form territory by introducing VeeValidate and saw how it can be used to define, handle, and validate our forms.

In the next chapter, we will take a step back from coding and learn how to investigate and solve issues by introducing debugging techniques and using the exquisite Vue debugger browser extension.

Part 4:
Conclusion and
Further Resources

In this final part of the book, we are going to explore reading materials, resources, and topics that will take your knowledge of the Vue.js framework further. In this part, we are also going to learn how to debug our application.

This part contains the following chapters:

- *Chapter 13, Unveiling Application Issues with the Vue Devtools*
- *Chapter 14, Advanced Resources for Future Reading*

13

Unveiling Application Issues with the Vue Devtools

If there is one aspect in which Vue.js is clearly the winner over other frameworks and libraries, it is its attention to the **Development Experience** (**DX**). Since the beginning, Vue.js has focused on providing great experience to its developers and, this reached its peak a few years ago with the creation of the Vue **Devtools** by Guillaume Chau.

The Vue Devtools, which is available as a Firefox and Chrome extension, or a standalone Electron app, has been at the core of the Vue.js DX since its creation.

Recently, Vue Devtools has been making headlines for its ability to provide great insight for **Nuxt.js** (Vue.js's meta-framework), helping developers comprehend the complexity of full stack JavaScript applications with simple and intuitive UI.

We are going to start this chapter by learning how to install and use the extension within our preferred browser; we will then learn its layout by understanding how each piece of the extension works and how they fit together. We will then examine each individual section, Components, Timeline, Pinia, and vue-router, to provide you with a complete understanding of the extension.

This chapter includes the following topics:

- Getting familiar with the Vue Devtools
- Deep diving into the Vue Devtools Timeline tab
- Analyzing additional data with the Vue Devtools plugin

By the end of the chapter, you will have a good understanding of the Vue Devtools and be able to use it in your day-to-day life. You will be able to use it to create new components, examine your application component tree, record user interactions in the Timeline, and finally, make use of package information such as Pinia and vue-router.

Getting familiar with the Vue Devtools

Even the most senior developers rely on debugging tools to help them develop high-quality and bug-free code (no code is really bug-free, but that is the aim when developing). The Vue Devtools goal is to provide quick insight into different parts of the Vue.js framework that can be used to help us complete our daily tasks.

It is possible to debug our application by placing `alert` and `console.log` or another preferred method within our code, but what if you could find all the information you need directly in the browser, using a very nice and clean UI? That is the Vue Devtools for you.

In this section, we are going to learn how to enable the Vue Devtools on our browser and learn the different sections of this extension.

In the course of this chapter, we are going to use the Vue Devtools Chrome extension, but the UI and features should match other available sources, such as Firefox and the Electron app.

To get started, we need to install the application on our browser; this can be done by searching for `Vue.js devtools` in your browser extension store. If you use Chrome, this can be found at `https://chromewebstore.google.com/category/extensions`.

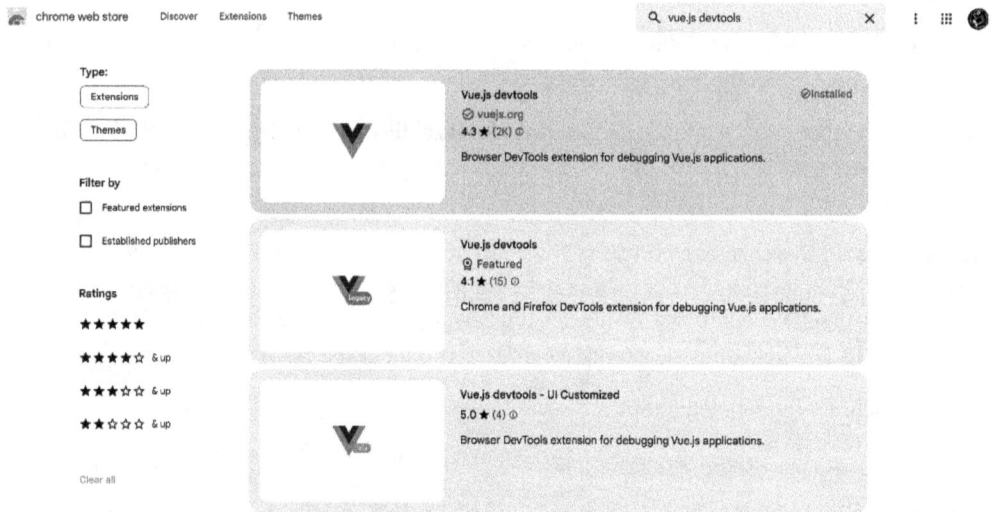

Figure 13.1: The Chrome extension store

There are a couple of extensions with the name **Vue.js devtools**, but you want to install the official one, supported and developed by the core team. This is defined by the **vuejs.org** checkmark.

After a quick installation and a browser restart, you should now have full access to the extension. In fact, the extension does not need any configuration, and it is good to go out of the box.

To test whether the extension is working, we can access our Companion App site at `http://localhost:5173/` and check the Vue.js extension icon. This is grayed out if the extension is not currently working, or colored if the Vue Devtools is available and in action.

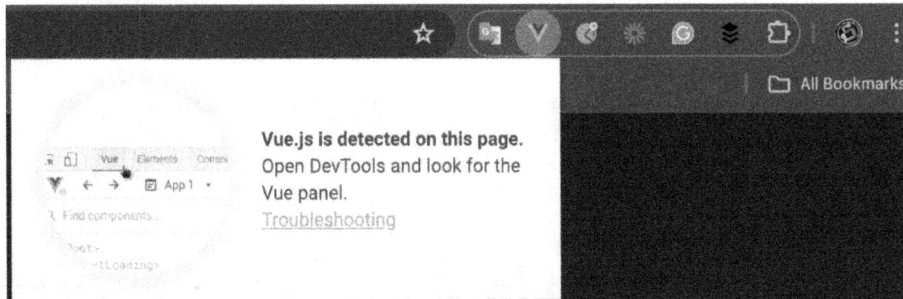

Figure 13.2: The Vue Devtools extension

Clicking on the icon will confirm that the extension was able to find Vue.js and enable the Vue Devtools, as shown in *Figure 13.2*. The extension lives within Chrome DevTools as a new tab. This tab should be available automatically as the last tab available within the DevTools navigation bar under the name of **Vue**.

Figure 13.3: Chrome DevTools navigation

> **What is Chrome DevTools?**
>
> If you do not know what Chrome DevTools is or have never used it before, I suggest you take a look at the official documentation at `https://developer.chrome.com/docs/devtools` and start to learn about all the different ways that this tool can be used.

The Vue Devtools automatically listens for a Vue.js application and is available immediately for all Vue.js sites running in development mode. This is a very important point, as the extension will not work on a website built for production. If you tried to access a production Vue.js site (such as the Vue.js official site), the extension would load but be inactive.

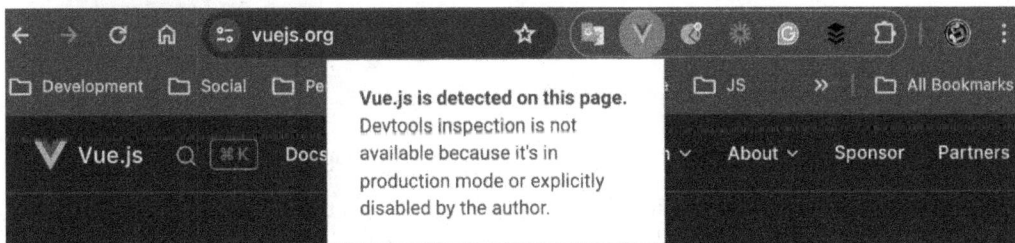

Figure 13.4: The Vue Devtools extension popup when accessing a production-built site

It is now time to return to the Companion App, open the Chrome DevTools, and click on the **Vue** tab to start to learn what this extension has to offer.

In this section, we are going to introduce the main parts of the extension, but as we will cover later in the chapter, the extension automatically expands to provide more information about different packages, such as Pinia or vue-router.

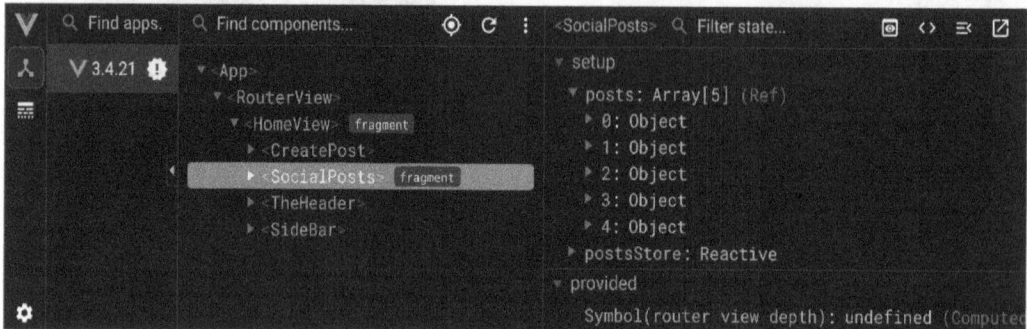

Figure 13.5: the Vue Devtools

The Vue Devtools can be divided into four main parts. We will briefly introduce you to all of them, starting from the left and moving toward the right:

- **The main navigation**: This is positioned on the left-hand side as a vertical menu. This currently holds two entities– Components and the Timeline.

- **The apps list**: Vue.js 3 allows you to have multiple applications on the same site, and the Vue. Devtools allows us to debug each of them easily.

- **The main content**: This is the area that includes the selected tool. In the case of *Figure 13.5*, we selected the Components, so this section displays a component tree.

- **The information panel**: This section is used to display specific details for the tool currently in use. In our example, we display details of the <SocialPost> component.

It is now time to start and learn the individual section of the extension. We will first start with the default sections, Components and the Timeline, and then move into add-on plugins such as Pinia and vue-router.

Debugging components in your Vue Devtools

In this section, we are going to see what Vue Devtools has to offer to debug and develop our components.

Having a tool such as Vue Devtools while developing your components can really help you improve your skills. Vue Devtools help surface information such as component properties, events, and other information that can sometimes be hidden or hard to grasp without a visual clue.

The main goal of the Component section of the extension, which can be reached by clicking the top icon in the main navigation, is to provide us with full visibility of the component tree and detailed information on each component.

Let's start by looking into the component tree and what information it provides us:

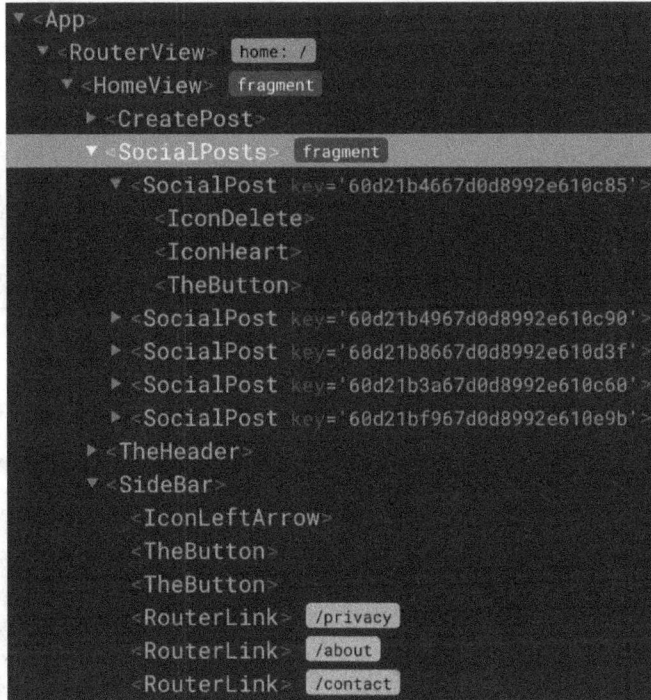

Figure 13.6: The component tree

The component tree provided by the extension is very similar to the DOM tree offered by browse developers' tools, with the main difference being that the DOM tree shows DOM nodes, while the Vue Devtools is made up of Vue components.

In *Figure 13.6*, we can see the relationship between the parent and children components. In fact, we can see that `<SocialPosts>` has five `<SocialPost>` children and that `<SocialPost>` has three children.

The tree also shows important information, such as the route used by `<RouterView>`, defined by the home : / purple pill, the unique key of each individual `<SocialPost>`, and the URL defined within each `<RouterLink>`.

Being able to visualize the components that are rendered on a page is important, but being able to highlight them is even better. You may not see this as important initially, but being able to select a component in the tree and see it on screen is a very useful feature.

There are two ways to highlight a component. You can click on the component directly on the tree or you can select the component in the page by enabling **Select a component on the page**, as shown in *Figure 13.7*. This last feature can either be enabled by clicking the icon or by using the keyboard shortcut, hinted by the popup shown when hovering over the icon.

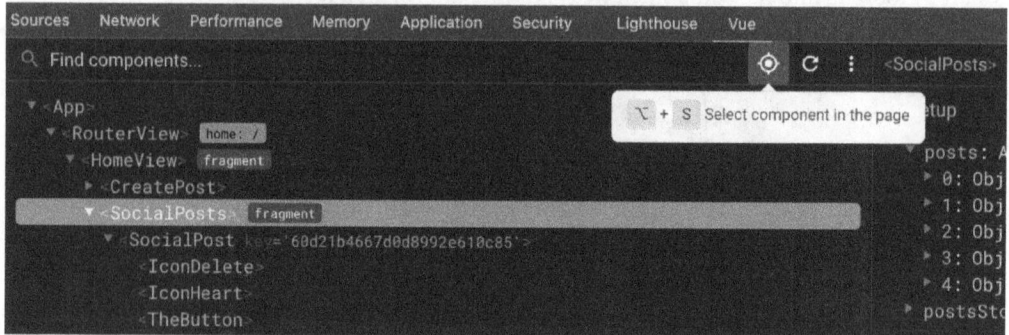

Figure 13.7: The Selecting a component in the page button

When a component is selected, it will be highlighted on screen with a green background.

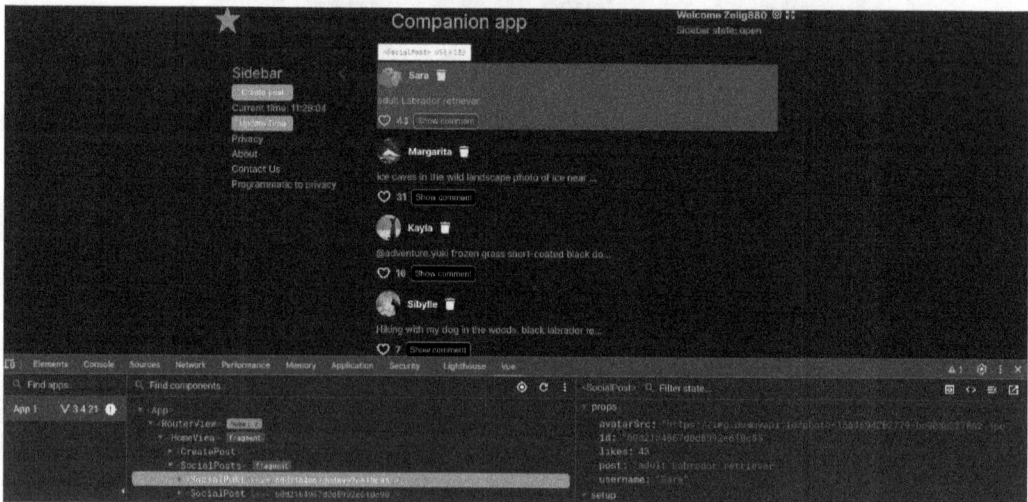

Figure 13.8: A social post component highlighted by Vue Devtools

Now that we can read the tree and select individual components, it is time to dive deeper into the extension and see what details are displayed for each component.

In fact, selecting a component not only provides you with a visual clue about the component by highlighting it in green as displayed in *Figure 13.8*, but also exposes a list of internal information about it.

The sidebar includes all the information available within the component scope and is the equivalent of being able to print everything that is available within the script setup. It includes basic information such as props, Refs, and Reactive and also more advanced features, such as Pinia store dependencies and router info.

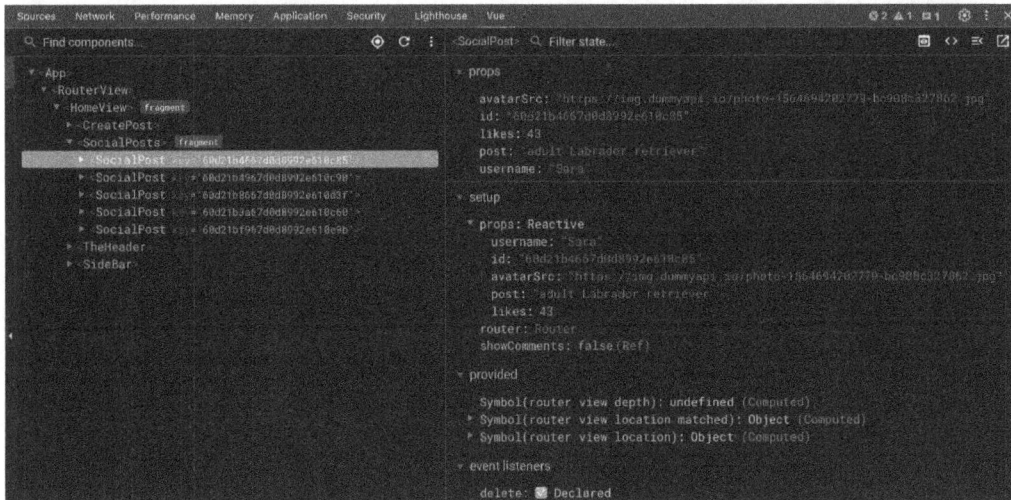

Figure 13.9: A social post's detailed information displayed by Chrome DevTools

Having access to internal information of each individual component on the screen is the most important and powerful feature of the Vue Devtools. Being able to quickly see what the properties that a parent is sending its child are, or the current value of a specific Ref will save you countless hours of debugging. Furthermore, seeing the component state will also help you better understand the Vue.js framework. When I first started to learn Vue, I used the Vue Devtools to understand how components worked and discover the connections between props, internal data, and life cycles, and I have recommended it to all my mentees ever since.

Use cases for the Vue Devtools information panel

In this section, we are going to explore a couple of use cases in which we would use the Vue Devtools information panel. As previously mentioned, the information panel includes lots of useful resources, but they are of no use unless we know how and when to use them.

Analyzing dynamically loaded data

In this first scenario, we are going to consider a case in which a developer would be required to debug and understand what data is returned by an API. In the case of our Companion App, this situation could have arisen when developing **social posts**. In fact, being able to see the data structure and value would have helped the development of that component.

We can see the complete information retrieved by the API as the `posts` array in `<SocialPosts>`, as displayed in *Figure 13.10.*

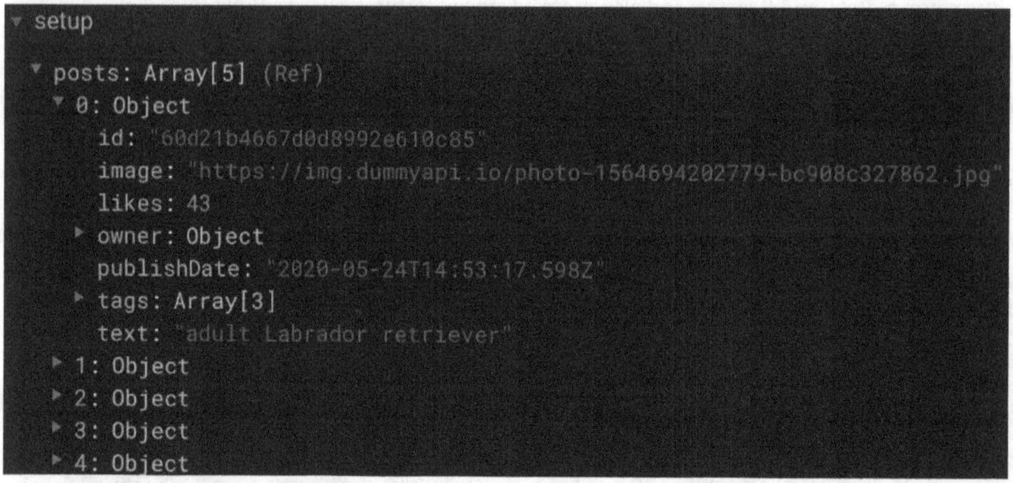

```
▼ setup
  ▼ posts: Array[5] (Ref)
    ▼ 0: Object
        id: "60d21b4667d0d8992e610c85"
        image: "https://img.dummyapi.io/photo-1564694202779-bc908c327862.jpg"
        likes: 43
      ▶ owner: Object
        publishDate: "2020-05-24T14:53:17.598Z"
      ▶ tags: Array[3]
        text: "adult Labrador retriever"
    ▶ 1: Object
    ▶ 2: Object
    ▶ 3: Object
    ▶ 4: Object
```

Figure 13.10: Information about the posts shown in the information panel of Vue Devtools

Modifying Refs and Reactive data

JavaScript has gained popularity because of the immense possibility to add interactivity to our web pages. Unfortunately, interactivity is not always easy to achieve, and it needs many tries before the logic usually works as expected.

While developing a Vue.js component, you will probably find yourself in a situation in which you will have to test different scenarios that would require data to be in a specific state. This can sometimes be very time-consuming or complicated to achieve.

A very common scenario would be the need to develop an "error" screen or a "a thank you page" displayed after an ecommerce purchase. To develop this component, developers would need to reproduce multiple steps to reach the desired component state, making the act of developing the component very slow. Luckily for us, with the help of the Vue Devtools, reproducing the states required to develop this component can easily be achieved.

In fact, Vue Devtools allows you to modify Refs and Reactive data on the fly. This can be done directly in the information panel, as shown in *Figure 13.11.*

> **You can modify props too**
>
> With the default setting, the only data that can be modified are Refs and Reactive, but there is a setting that allows you to also modify Props. I would personally not suggest enabling this, as it can produce unexpected consequences, but knowing of its existence can be helpful as a last resource.

```
post: "Two boys hug their dogs in a leaf pile in the fall...
likes: 28
router: Router
                              Quick edit
showComments: false(Ref) ✏ ☐ ⋮
```

Figure 13.11: A Vue Vue Devtools quick edit

Monitoring changes

Modifying the data within a component is not the only requirement to develop complex components. In fact, another important feature is the ability to see in real time the values of component data. For example, you may want to ensure that your toggle button works successfully but that a Ref changes from false to true, that a computed property changes its value as expected, or finally, that a "reset" logic works properly by resetting all data to its initial state. The possibilities are endless.

When working with the Vue Devtools, the data displayed in the information panel will change automatically in real time when modified. This allows us to interact with the application and see the changes that our interaction yields.

In this section, we introduced the Vue Devtools by learning how to install and enable the extension. We then defined all the different areas of the Vue Devtools and learned how to use it to highlight and read the Vue component tree. Finally, we learned how to dive into a component by accessing the information panel. In this panel, we learned how to see data, monitor changes, and even make changes on the fly.

In the next section, we are going to introduce a more advanced section of the DevTool, the Timeline section.

Deep dive into the Vue Devtools Timeline tab

In the previous section, we introduced how to use the Component section of the Vue Devtools to analyze and develop our Vue components. In this section, we will continue exploring the main navigation to introduce the next available section – the Timeline.

The Timeline section is home to tools such as event and performance monitoring that help make sense of the application, giving us a glimpse into the framework engine.

This section is probably more for advanced use cases, but being able to know what the tools offer is always beneficial, even if they are not used in your day-to-day activities.

The Timeline tab can be accessed from the main navigation on the left-hand side of the extension, just below the Component tab that we have been using so far.

The Timeline panel includes three sections. The first is the layers section, which displays all the different layers on which we are going to collect and display information. The second is the Timeline itself. This is provided in two different ways, as an actual Timeline or in a table format. Finally, we have the information panel, which is very similar to the one provided in the Component section.

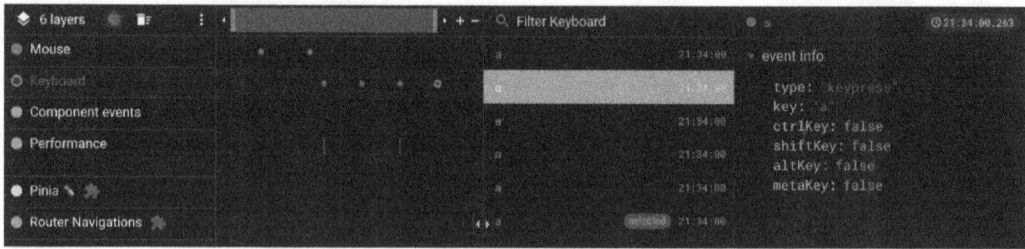

Figure 13.12: The Vue Devtools Timeline panel

The layer section includes all the different information that the Timeline can provide. The layers offered by the Vue Devtools can differ between applications, as it can also include an installed package, like the one shown in *Figure 13.12*, where data for Pinia and Router is also available within the layer's menu.

Layers can produce lots of data (as we will see later in the section), and it is common to hide most layers and just keep the ones that are relevant to your specific task.

Enabling the Timeline is resource-intensive, and it does not run automatically all the time. To be able to enable the Timeline, we would need to click the record button next to the layer's header.

Being able to start and stop recording will not only help us save some battery life; it will also ensure that the data displayed in the Timeline is as compact as possible. So, for example, if you want to analyze what happens when a user creates a post, you will enable it, create the post, and then stop the recording. Doing so will ensure that just the events you are interested in are shown on the Timeline.

Debugging a post removal

To better understand the Timeline, we are going to record and analyze the information provided by the Timeline when removing a post and try to determine how this can be used for future debugging purposes.

To be able to analyze the output of the Vue Devtools Timeline section, I am going to perform the following steps within our app. These steps will produce a report that we can use to understand what information the Timeline provides:

1. **Access the Timeline tab**: Open DevTools and click on the Timeline tab on the main navigation.

2. **Select the layers**: We will only use the **Component events**, **Performance**, and **Pinia** layers. Select them from the layer's menu.

3. **Start recording**: Now, we are ready to record our actions. Start the recording by clicking the record button.

4. **Perform actions within the Companion App**: While the Timeline is recording, let's go back to our application and start to perform some actions. In my example, I have clicked the `"delete"` icon of one of the posts shown on the screen.

5. **Stop the recording**: Click on the record button again to stop the recording.

After the preceding steps, the Timeline should display something like what is shown in *Figure 13.13*.

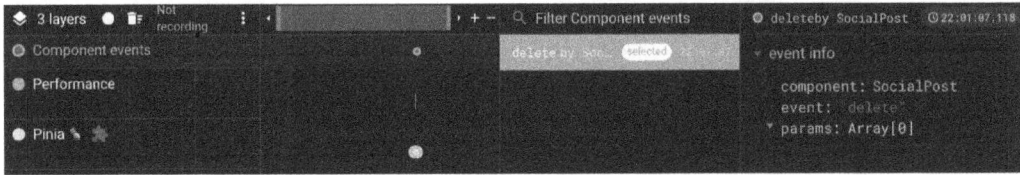

Figure 13.13: The Timeline displaying information after deleting a post

Let's see what information is shown after our recorded activity. Before checking the output of the recorded session, let's go back to the application code and discover what action we expect from the aforementioned action. The code shows that when clicking the delete icon within the `<SocialPost>` component, it should emit an event called `"delete"` that is then used in `<SocialPosts>` to trigger a Pinia action, subsequently deleting the post from the store.

If we go back to our recorded activity, we can see that the Timeline shows some activity in all three layers. The first layer, **Component events**, shows that an event has been triggered. This event is named `"delete"`, and it has been triggered by `<SocialPost>`, just as we expected. The information panel shows detailed information about the event, such as params, but in this case, we have none and the array is empty.

Next, we are going to see what the **Pinia** layer shows:

Figure 13.14: Pinia's timeline event

After selecting the **Pinia** layer, we are able to see three different entries. The first is the start of the `removePost` action, with detailed information displayed in the information panel. Next, we have a mutation; this is called from `removePost` and is the actual deletion of the post from the `posts` array. Then, we have the end of `removePost` with the `removePost end` event. Overall, the action took 0.8 ms to complete. So far, we have been able to analyze whether an event was triggered and follow the steps of a Pinia action.

> **Keep track of your actions**
>
> Our `removePost` action was very fast and completed in less than 1 ms. This is not always the case, as some actions may include complicated code or external operations that may delay its execution. Using the Timeline can help you debug and fix slow actions.

Finally, we are going to check what information is shown in the performance layer. Just like the other two layers, when selecting this layer, we are presented with individual events and information about the events.

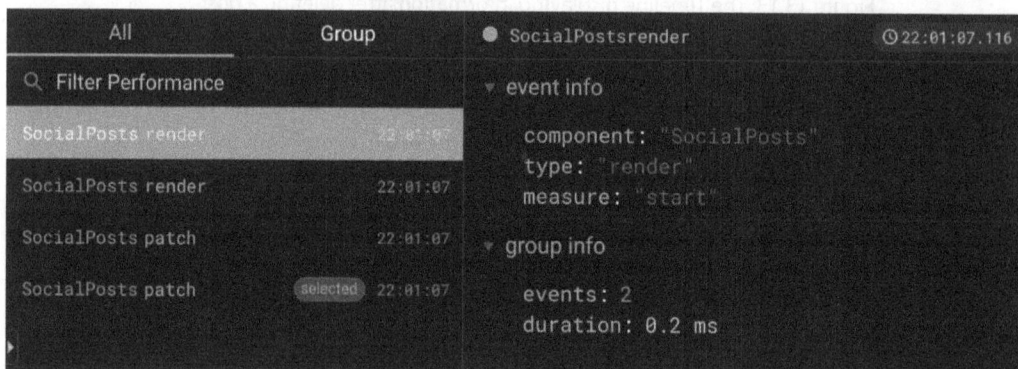

Figure 13.15: The performance Timeline layer

The performance layer is used for advanced use cases, and it is usually accessed to unearth performance-related issues or to highlight an incorrectly configured code base that forces you to re-render one or more components.

This layer displays when a specific component is rendered or updated. In this scenario, it just recorded four events, but this can be extremely complex for larger applications.

Being able to analyze the rendering time and how many times a component renders can be very useful for applications that suffer from performance issues. Incorrect use of **computed properties** or **watch** can result in the unnecessary re-rendering of components, and the performance Timeline is the best tool to help you debug and fix performance bottlenecks.

In this section, we have learned how to use the Timeline section of Vue Devtools. We have defined its structure, recorded a sample test to see how our application performs and learned what information is shown by the individual layers.

In the next and last section, we are going to see how Vue Devtools can be expanded with the help of custom plugins.

Analyzing additional data with Vue Devtools plugins

You have probably noticed from the previous section that Vue Devtools not only includes information regarding the core framework but also exposes extra information, such as the Pinia store and vue-router.

This happened automatically, as we did not have to do anything extra to expand the Vue Devtools capabilities. In fact, the plugins are actually part of the packages that we installed in the application, so installing a package can sometimes result in an additional feature being displayed in our Vue Devtools.

Any plugin offers something different. In our case, both Pinia and vue-router adds a layer in the Timeline view, information in the component details panel, and finally, an extra tab on the main navigation.

In this section, we are going to deep dive into the two available plugins – Pinia and vue-router.

The Pinia Vue Devtools plugin

The first extension plugin that we are going to analyze is the one provided by the Pinia store.

The Pinia store, distinguished by its iconic Pineapple logo, can be accessed from the main navigation on the left-hand side of the extension.

This plugin will feel somewhat familiar, as it has a similar layout to the Component section.

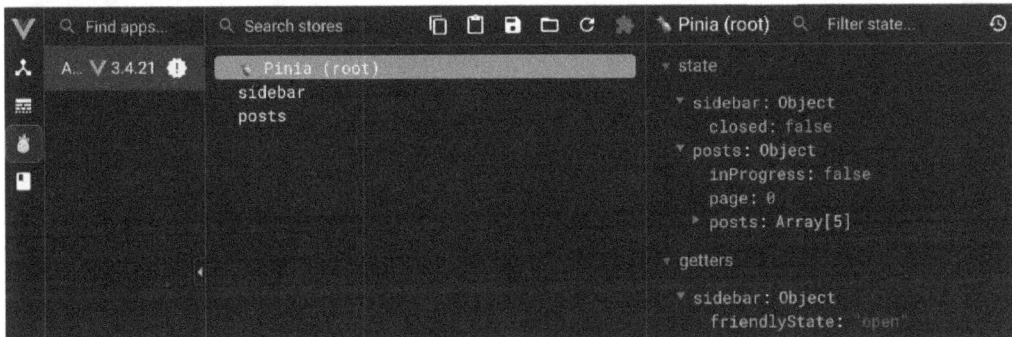

Figure 13.16: The Pinia Vue Devtools plugin

In the Pinia extension, we are able to inspect our stores. The plugin offers the ability to see all stores together, by selecting **Pinia (root)** or the individual store available within our state management.

The data shown in the information panel is divided by state and getters, and just like the Component section, it can be modified on the fly.

Just like most of the features we have explored so far, being able to quickly review the current state of your stores and modify their values is extremely valuable. Having access to this information on the fly will save you countless hours of development.

The vue-router Vue Devtools plugin

As you may remember, back in *Chapter 10*, where we introduced the router within the Companion App, we had to create a set of rules that would be used by the router to decide which route to display to our users.

The vue-router plugin helps visualize and examine these rules directly in your browser. The plugin provides a complete list of the routes defined with their full information, including names, regex matches, and required keys.

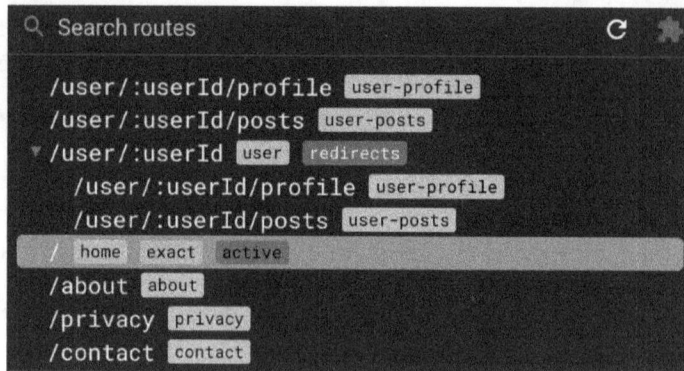

Figure 13.17: The vue-router rules list

The list displayed in *Figure 13.17* shows the following information:

- The URL defined for a specific router (`/about`).

- The name of the route shown as a blue pill (`user-profile`).

- The current active route, defined by a blue pill with the `active` label.

- Extra information such as `redirect` and `exact`. That signifies respectively the presence of a redirect rule (`/user/:userId`) and the notion that the currently active route matches perfectly to the rule match regex.

Now that we have defined all the information available within the vue-router rules list, it is time to see what information is provided for each individual router rule. Let's click `/user/:userId` to see what it shows:

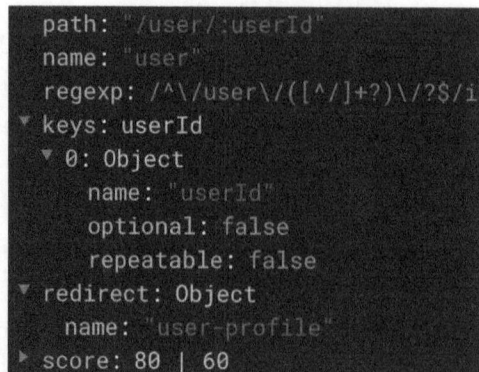

Figure 13.18: Route information from the vue-router Vue Devtools plugin

The plugin displays all the data that is available within the routes array, such as `path`, `name`, and `redirect`. Furthermore, the plugin exposes more advanced data, such as the regex used by the package to decide the correct route to display, details information on the keys, such as `optional` and `repeatable`, and, last but not least, the `score` value of each entry within the path.

During your career, you are probably not going to spend lots of time in this area, but when you do so, it will have everything you need to solve your problems.

Summary

The Vue ecosystem prides itself on offering one of the best DX available within the whole industry, and Vue Devtools is probably the one to blame for this.

Detailed information, performance metrics, and an automatic plugin extension make the extension a must-have for all Vue developers. The most important take from this chapter is for you to install the extension locally and start to use it during your projects.

In this chapter, we learned how to install and navigate Vue Devtools. We covered the individual default sections offered – Components and the Timelin. Finally, we ended the chapter by introducing the power of the package plugin within the Vue Devtools, by showing the additional information that Pinia and vue-router provide within the debugging extension.

In the next and final chapter, we are going to touch upon future learnings and resources that you can use to continue your journey to becoming a Vue developer.

14
Advanced Resources for Future Reading

If you have reached this chapter, there is a high chance I will meet you soon within the tech community as a successful Vue.js developer. This book aimed to provide you with enough knowledge of the framework to give you a head start in your career, and the fact that you persevered until the end of the book highlights your determination, which is a key skill to have in the world of tech.

In this chapter, we will share useful resources, introduce advanced Vue.js topics, and discuss ways to improve your skills as a Vue.js developer.

This chapter is divided into the following sections:

- Available tools and resources
- What is left for Vue.js
- Contribute back to the community
- Let's look back at what we have achieved

By the end of this chapter, you should have gained a good understanding of what you have achieved, and you will have learned about the resources available for you to continue your learning beyond this book. Within this chapter, we are also going to introduce further topics that were too advanced to cover at this stage, but they are useful as learning materials to continue your training.

Available tools and resources

Vue.js and its ecosystem, just like other frontend technologies, are affected by continuous changes and updates that help it stay relevant and improve its features. In this section, we will learn how to keep informed with the latest news within the Vue.js community.

We will introduce newsletters, websites, community members, and more. Keeping informed within a fast-paced industry such as JavaScript is vital for your career. In fact, during your day-to-day job, you will usually work on the current or previous versions of software or library, and this will not expose you to the latest features or improvements included in future releases.

Not being able to use upcoming releases in production is normal, and it is vital for you to find a different way to keep yourself informed about all the new trends and features. To do so, we are going to discuss different resources that will help you keep up to date.

Vue.js documentation

We could not start this section without mentioning the official documentation. Vue.js' documentation (`https://vuejs.org/`) is always up to date and well defined and it should be your go-to place to see information about new features or changes.

The website offers lots of interesting areas, such as the example pages and the tutorial, but the following links should be bookmarked and used during your day-to-day work:

- **Vue Blog** (`https://blog.vuejs.org/`): Access the latest news about new releases and upcoming changes. This is the first place where you will hear news about the content and information of new releases.

- **The Vue.js API cheatsheet** (`https://vuejs.org/api/`): A one-pager with a shortcut to all the Vue.js APIs. This will be invaluable at the start of your career. As a developer, you are not expected to remember everything, but to know what exists and how to easily access it, and this page just does that. It provides you with the information you need with just one click!

- **Vue Playground** (`https://play.vuejs.org/`): This may not provide you with new info, but it is extremely useful for trying new things without needing to set up a full environment. The link is also sharable, with no need to sign up or authenticate.

Vue Discord (`https://discord.com/invite/vue`): What better way to learn than by being part of the actual Vue.js community? You can join the official Discord chat and join over 10,000 people, including core members and library maintainers. The Discord channel includes many chats that cover the main framework and its libraries. The chats are very active and are moderated by the actual Vue.js core team.

Newsletter

Vue.js' official documentation is a really great place to find any information that you may need or require, but it will require you to go to the site to get your information. In today's world, we expect information to be delivered directly to our inbox, and what better way to receive information than by subscribing to some newsletters?

Here is a list of Vue.js-specific newsletters that I suggest you subscribe to:

- **Weekly Vue News**: `https://weekly-vue.news/`
- **Vue.js Developers**: `https://vuejsdevelopers.com/newsletter/`

Both newsletters provide great content and up-to-date information on the Vue.js ecosystem, and they are free. These newsletters are delivered once a month and are perfect for a weekend read.

If you have time to read each individual post shared you should surely do so, but if you do not, I suggest you take the time to at least scan through the names of the posts, as they can provide you with important information, such as a major release or a breaking change.

Community members

It is quite hard to be a developer without being part of a social media platform. I am personally very active on X (`x.com`) under the username of `@zelig880`. In this section, I am going to highlight a couple of important community members that you should personally follow on your social media of choice.

These people are what makes Vue.js unique. They help the community grow by sharing their knowledge at conferences, being active members of the core repositories, and by being active library maintainers.

I am going to put each X username next to the name to make your search on the web easy, because some people may have very common names:

- **Evan You** (`@youyuxi`): Evan is the Vue.js community member of excellence, not only because he is the creator of the framework, but also because of his active involvement within the community. Evan is extremely active at conferences and on social media, and he is the first to share upcoming news.
- **Anthony Fu** (`@antfu7`): Anthony is behind many libraries in Vue.js, including **VueUse** and **Vitest**. Anthony not only creates amazing libraries, but he also likes to experiment with new things, so his timeline is always full of nice ideas and creations.
- **Eduardo** (`@posva`): Eduardo is the creator of Pinia and an active member of the Vue ecosystem. Eduardo travels the world to speak at various Vue.js conferences around the globe.
- **Daniel Roe** (`@danielcroe`): Daniel is a Nuxt.js core team member and a very active conference speaker. Daniel also likes to stream his work. Watching an experienced person such as Daniel trying to complete a task and, more importantly, debug issues online is a fantastic way to learn.
- **Jessica Sachs** (`@_jessicasachs`): Jessica has worked at many companies, including Cypress, and is an active conference speaker.

I could continue to add more amazing members of the community, but I think this list is great for you to get started. Following these people will provide you with good insights and updates within the framework ecosystem.

Request For Comments – RFC

To complete this section, we will share something a little more advanced, but it still helps you to feel part of the community: **Request for Comments** (**RFCs**). Vue.js is the only major framework not being backed by a tech giant. Being independent gives it the flexibility to make choices based on the community and not to please shareholders (I know that most frameworks are not for profit, but their decisions are still made behind closed doors).

If you were to ask Evan You if he is the owner of Vue.js, he would say no, and he would clarify that Vue is owned by its community. After years in the ecosystem, I can confirm that this is not only something that he says, but it is the way he and the rest of the core members build the framework.

With all its major releases, Vue.js actively uses RFCs to share ideas and changes in its core engine. This repository has seen many conversations, some extremely lively, but most importantly has shown the openness of the core team to help.

The repository, including all the RFCs, can be found at `https://github.com/vuejs/rfcs`.

I do not expect you to look at the repository daily, but it is very beneficial to look at the comments and discussions when a new topic is shared in newsletters or by a community member.

For most developers, the conversation and code shared in these RFCs can be quite advanced, but reading through the comments can help you grasp how people use the framework and help you grow in the process.

This was the last topic for this section, in which we shared different ways in which you can keep active within the Vue.js ecosystem, we discussed the usefulness of the official documentation, discussed how to receive information by subscribing to up-to-date newsletters. We then had a section on community members and introduced a few important people who should be followed to be the first to hear about news and updates on the framework. Lastly, we discussed RFCs and learned how they can give you a glimpse into the future of the framework.

In the next section, we are going to learn what else is left for us to learn in Vue.js and what other topics you may encounter in your journey as a Vue.js developer.

What is left for Vue.js?

In this book, we have covered many different topics and learned enough about Vue.js to be able to handle your future tasks or projects. But we have just learned about the core features of the framework, and there are still some advanced topics that you may encounter in the future of your development journey.

These topics are no less important than the ones you have already covered, but they are secondary because they come after the basics have been learned and mastered. In this section, we are not going into the details of each of the topics, but just give a short introduction.

The idea behind this section is to provide you with a basic understanding of many topics that you may have to learn in the future as you progress your career as a Vue.js developer. Some of these topics are quite advanced, while others may just be needed in special use cases.

We are going to break the topics into sections: Miscellaneous, Core, Pinia, and Vue-router features. Let's start first by looking at what topics we would have had to learn to make our Companion App production-ready.

Miscellaneous

The following topics would have been necessary for the application to be in a production-ready state, but are too advanced for someone who is just starting to work on the framework.

The list includes big topics, such as authentication, that could have entire books devoted to them. The reason why we are going to cover these topics is not to make you go and learn about them right away, but for you to be aware of their existence and learn about them when the time comes for you to implement such features.

Let's see what we have missed in our application:

- **Authentication**: An application such as the Companion App could never be completed until we have set up authentication. Luckily for us, there are many services, such as **Auth0**, **Supabase**, and **AWS Cognito**. These tools have great documentation and starter guides. Using an authentication provider is great not only because it saves you development time, but also because hosting usernames and passwords in your servers is a security risk and a complicated task that you should avoid at the start of your career.

- **Environment variables**: In our application, we have set variables such as tokens and URLs directly in the application, but this is not correct. What we should have done was use environment variables to store these values securely away from the repository. Vite provides a simple way to define environment variables (`https://vitejs.dev/guide/env-and-mode`).

- **Logging/bug reporting**: An application should never go into production without good logging. Creating a bug-free application is a myth, and there is always a possibility that your user will run into trouble. Most of today's hosting environments provide logging out of the box, but I prefer to use tools such as **Sentry** that help you quickly find the issues in your application, saving you time and money.

- **CI/CD**: This term stands for **Continuous Integration/Continuous Delivery**. It aims to streamline and accelerate the software development life cycle. Before promoting an application to a production state, you may want to set up an automatic pipeline that quickly delivers your code into production. Luckily for you, there are also services such as **Netlify**, **Vercel**, and **AWS amplify** that offer ready solutions for this.

- **Component library**: At first glance, creating components may seem like a simple task, but this is not true. A fully fledged component library can take thousands of hours and it is not feasible for most developers. Using an existing component library such as **Vuetify** (`https://vuetifyjs.com/`) or **Quasar** (`https://quasar.dev/`) can speed up our development and help us focus on what matters the most: the application's logic. If you are searching for a specific component, you can search for it on vuecomponents.com, that is an active list of available components within the Vue.js ecosystem.

Core features

It is now time to cover Vue.js' core features that could be of use in future tasks. Just like everything else within the Vue.js ecosystem, these core features are well documented and can be learned from the official Vue.js documentation:

- **Composables**: These are functions that use Vue.js APIs to encapsulate stateful logic. Composables are used to reduce duplication and help you reuse code. Technically, composables are just `<script>` tags of a **SFC (Single File Components)**, but the reasoning behind it and the way you use it may require some further learning. The best way to learn about composables is to find some composable that you can add and use in your application, and there's no better place than **VueUse**. This is a list of hundreds of composables created and maintained by Anthony Fu and many other active contributors. Adding this to your project will not only speed up your development, but also help you fully understand how composables work and what problems they solve. As an added bonus, VueUse is open source and their code is readily available from the website.

- **Components**: You may be confused by what the word component means in this context, and that's OK. The heading component in this case refers to a "meta component" provided by Vue.js that allows you to render dynamic components or elements. You may use `<component :is="componentName" />` to conditionally render a table or a list, or to change an `<input>` to a `<textarea>`, depending on the user preference. Just like other topics in this section, this built-in component solves a very specific problem.

- **Scoped Slot**: Scoped slot is a very advanced topic. It's rare you'll ever need it except in specific situations. Scoped slots solve a very specific problem, but unfortunately not the easiest to use or learn. Remember their existence, and if you are ever trying to achieve something with slots and cannot do so, search for them as they will probably be the solution to your problem! But as I said, avoid them for now and ask a senior developer to help you if you need to .

- **Transition**: It is quite common for websites to include some animation. This may be a little change on a button, or an effect shown between two elements' states. Even if transitions and animations can be achieved using plain CSS, Vue.js provides us a component called `<Transition>` that helps us mix the power of CSS animation with the flexibility of Vue.js templating power such as `v-if`.

- **Teleport**: Do not worry, we are not going to start talking about science fiction. Teleport is another built-in component from Vue.js that allows you to "teleport" a Vue.js component into a different node of the DOM that lives outside of the Vue.js application. This component solves specific problems where the logical placement of a component may not match its visual placement. A common use case for this is a dialog box, where the modal button and component would live within the same component (logical placement), while the visual of the dialog itself should not be nested within the DOM but higher up in the tree.

The list of advanced components and techniques that we just shared is just the tip of the iceberg. Vue.js has still lots more to offer and many advanced topics that you will discover during your career.

Pinia's features

In *Chapter 11*, we introduced state management using the official library, Pinia. In this section, we will share a couple of advanced topics that we have not seen so far in this book.

Just like we previously mentioned, these are good topics to know, but they are not essential to learn at the very start of your journey. Let's see what else Pinia has to offer:

- `$Patch`: The biggest feature of Vue.js is its reactivity system. As we saw in the previous chapter, Vue's reactivity is also extended with Pinia, which provides a store that is sharp to respond to changes. To ensure that your application is performant and that the reactivity does not turn into a negative effect, Pinia provides a method called `store.$patch`. This is used to modify multiple store entries at the same time. This method, on top of accepting a partial state object, also accepts a method that can be used to update complex parts of the store, such as arrays or nested objects, and ensure it all happens in a single state change.

- `$reset()`: Another method available within the Pinia store is called `$reset()`. This method allows you to restore the state of a store to its initial state. This is extremely useful for dynamic forms and other interactivity that would require you to clean up the store after the complete execution of an action. Calling this method just returns the state object to its initial value.

- `$subscribe`: Of the Pinia methods that we are covering in this section, this is probably the most advanced one. Pinia exposes two different methods to subscribe to part of the store. The first is the `$subscribe` method. This allows you to watch for a Pinia store change. The second is called `$onAction`, and it is used to listen to Pinia's actions. Subscribing to a Pinia store with these two methods is not an everyday task, and it is usually used to solve specific problems.

After Pinia, it is time to look at the vue-router and complete the last part of this section.

Vue-router features

Just like Pinia, we introduced vue-router back in *Chapter 10* and learned how to implement the basic router features within our Companion App. In this section, just like we did in the previous sections, we are going to introduce advanced technique offered by the router library:

- **Lazy loading routes**: The first feature we are going to share is **lazy loading routes**. Vue-router allows you to specify a route to be loaded asynchronously. Loading routes asynchronously will improve the performance of our application by reducing the size of the JavaScript bundle our visitors have to download initially and loading the routes on demand.

- **Guards**: Router guards allow vue-router to "guard" user navigation by either redirecting or canceling navigation. Router guards are useful for enforcing role-based navigation or authentication. With router guards, you can run logic before or after each navigation, allowing you to take full control of what the user is allowed to see.

- **Active link**: Requiring the ability to show an active state within a main navigation or a site is common practice. Vue-router simplifies this by providing us with active link classes. These are automatically assigned to router links that match the current URL fully or partially.

- **Named Views**: This last feature is one of my favorites because it makes it possible to implement complex application layouts with very simple code. Until now, we have defined our `<RouterView>` using a single outlet, but with named views we can provide different named sections of our layout. We could use this to define a layout with a sidebar and a main body and swap the sidebar depending on the route we are in.

More information about the features we just discussed can be found in the official vue-router documentation at `https://router.vuejs.org/guide/`.

Contribute back to the community

Throughout this book, you have heard how amazing the Vue.js community is and seen first-hand the great work that the core team and the package maintainers do. The Vue.js framework is well documented and well maintained, the development experience is well defined, and the same applies to core-maintained packages such as vue-router and Pinia, but also external ones such as VeeValidate and many more within the ecosystem.

This level of consistency does not come for free, and there is a lot of work that goes on behind the scenes to ensure that the community feels accepted and supported. This work is at times done by developers who, due to the nature of open source, do not earn a wage for their contribution.

After reading this book, you have become an integrated part of the Vue.js community, and you should try your best to contribute back to the community and help maintain it. Doing so will not only support the ecosystem and future developers, but it will also help you grow.

Let's see the different ways in which a community member can get involved in the community:

- **Documentation**: If you are not new to open source, you have probably heard people asking for help with documentation many times before. This is mainly because it is truly the most important and easier action that can be taken by developer to contribute back to the open source community. As a new member of the Vue.js community and first-time user of the technology, you will be able to assess what documentation is well written and what may need to be improved. All Vue.js documentation is open source, so if you find something that needs fixing, create a PR.

- **Reproduce bug report**: When I first started to learn Vue.js, this was one of the main sources of learning material for me. In fact, there are plenty of bugs that are raised within the core framework and its libraries, and quite a few that are not accompanied by reproducible code or have an incorrect reproduction. Trying to reproduce a bug that has been shared is a very good way to learn more about Vue.js. These issues can be found in the main repositories with the label **need repro** (`https://github.com/vuejs/vue/labels/need%20repro`).

- **Host meetups**: There is no better way to help Vue.js grow than by hosting a local meetup. Hosting a meetup may not be for everyone, but if you have the right personality and interest, it can be extremely rewarding. You will not only gain popularity within the industry, but you will also learn a ton as you will meet other people within the Vue community and also hear about them during your events.

- **Be an active Discord member**: In the first section of this chapter, we mentioned Discord as a way to keep yourself informed within the community. In this section, we mention Discord again as a way for you to contribute to the community. You may not be able to help straight away, but as you gain some knowledge, you should try and help triage some of the questions and discussions shared in the chat.

- **Donate**: If you are starting your development career now, this may not be something that you can afford, but as you grow into your career you should consider it. Vue.js does not have a huge company backing it, and most of the core maintainers' wages are made up of small donations made by people like me and you. Every small donation helps.

No matter how you do it, what matters the most is that you try to give back to this amazing community and become an active part of it. In the next and last section of this chapter and book, we are going to summarize what we have achieved within the course of this book.

Let's look back at what we've achieved

You have reached the very end of this book, and you are ready to go and experience Vue.js on your own. When learning new technologies, like you did in this book, it does not come easy, and it is possible that you will need to go back and revisit some of the topics that we covered during the course of the book.

In this section, we are going to have a quick recap of what we have learned and what we can accomplish with it. Going over a topic multiple times will not only help you memorize it, but it will also support you in understanding its meaning and therefore allow you to apply it to different use cases.

All about Vue.js

We started the book by learning about Vue.js and, most importantly, about its reactivity system. Across different frameworks, Vue.js' reactivity is very powerful, and understanding it is a must if you want to improve your skills.

Next, we learned about Vue.js lifecycles. The framework works on a well-defined cycle, and learning about it is essential to enable you to create performant applications.

Finally, we completed the section about Vue.js by learning the structure of a Vue.js component also known as a **Single File Component** (**SFC**). We learned about its composition and used it extensively in the book.

From the basics and beyond

With the help of our Companion App, we have learned how to turn static HTML into a dynamic Vue.js component. We learned the basics of the framework by introducing string interpolation, Refs, and `Reactive`.

We then continued to increase the complexity of our application by introducing some of the Vue.js built-in directives. With features such as `v-if` and `v-for`, we were able to increase the amount of functionality offered by our components, still being able to write elegant components.

Last, we introduced `computed` and `methods`. These gave us the ability to add logic to our components.

With the knowledge learned so far, we could turn a very static website into a dynamic and powerful web application that would outclass simple static HTML sites.

From component to components

From *Chapter 6* onward, we started to expand our learning from a single component to multiple ones. We did so by introducing props and events. This allowed us to create components that could speak and depend on each other.

With these extra topics, we also had a chance to re-introduce and use Vue.js lifecycles, which we learned about at the start of the book.

To finish it off, we started to connect all the different topics we'd covered up to this stage by loading data asynchronously. To do so, we had to use methods, lifecycles, events, Refs, and props.

At the end of this section, which concluded with *Chapter 7*, we concluded the foundation of Vue.js.

Vue.js ecosystem

After the basics of Vue.js, it was time to move on and introduce important topics such as tests, routers, and state management and move away from the core framework.

In the chapters that followed the basic topics, we first learned how to write unit and E2E tests in our applications with Cypress and Vitest.

Next, we introduced two libraries maintained by the Vue core team members: Pinia and vue-router. With Pinia, we introduced the notion of state management and refactored our Companion App to use Pinia to support its state, while vue-router was used to implement a routing system and transform our single application into a complex application with multiple pages and nested routes.

After routes and state, it was time to add some user interactivity by learning how to manage forms. We did so by introducing the notion of two-way binding with V-Model and then discussed how using an external library such as VeeValidate could help us develop complex and well-structured forms.

Last, we stepped away from coding and learned how to use the Vue.js devtools extension to help us develop and debug our Vue.js applications.

Just the tip of the iceberg

Along with the books, we have discussed possible exercises and learning that you could undertake to improve your understanding of the various topics.

The information shared within the book and the exercises provided were just the tip of the iceberg. To fully understand how Vue.js works and be prolific with it, you will have to continue to practice and use what you have learned in this book.

What I suggest you do is keep practicing, maybe by setting specific daily involvement goals or by joining communities such as #100daysofcode. You can practice by creating very small proofs of concept using Vue.js Playground to test your knowledge or replicate a small part of existing applications with all their knowledge and interactivity to push your understanding. After you feel like you have gained enough understanding of the foundations of Vue.js and its ecosystem, it would be a good idea to try and clone an existing web application. This is an extremely useful practice that can help you focus on the practical part of learning Vue.js without the need to come up with an app idea from scratch. A few examples of applications that could be replicated are social media platforms, news platforms, and weather apps. They provide a good level of interactivity and offer free APIs and resources.

Summary

After 14 chapters aimed at improving your skills as a Vue.js developer, we have finally reached the end of this book. Completing this book is probably associated with the beginning of your career in Vue.js.

What you have learned in this book should be enough to start your career, but it is not going to be enough on its own. What is left for you now is true dedication and consistency in trying to apply everything you have learned here to a real-world project. As you do so, you may find yourself going back and revisiting a chapter or section or reading further to deepen your knowledge (always start your research with the Vue.js official documentation).

Vue.js is a fantastic framework, and I can assure you that you have made a good choice. As we mentioned time and time again, what makes Vue.js unique is not just its code and features, but the full ecosystem and community built around it.

Finding a good way to engage with the community will help you stay motivated and engaged, but, most importantly, interacting with other Vue.js developers can help you overcome issues and obstacles that you will face throughout your career.

All Vue.js core libraries and its ecosystem are built on top of the framework's core logic. This not only makes things familiar as you move between libraries, but it also means that mastering the framework's core foundation will help you understand the whole ecosystem. I have had the pleasure of meeting many Vue.js developers, and the trait that made the most senior developers stand out was their understanding of Vue.js to its core. This does not need you to learn the actual code base, but just try to really focus and understand the topics covered in the first two chapters, such as **reactivity** and **lifecycles**. Vue.js reactivity will help you understand when things update and why, while the lifecycles will provide an understanding of how the framework re-renders and the flow that it follows when doing so. I usually like to compare development to a dance. There are different steps, and you may not know them all at the very start, but what matters the most is that you know how to listen to music and move with it. The rest will come as you keep practicing. Learning the foundation of Vue.js is the same as being able to listen to music, and knowing it well will make the rest of your journey easier.

At this stage, I can just wish you all the best in your future and hope that Vue.js turns out to be a good choice for your career.

Index

‹packt›

packtpub.com

Subscribe to our online digital library for full access to over 7,000 books and videos, as well as industry leading tools to help you plan your personal development and advance your career. For more information, please visit our website.

Why subscribe?

- Spend less time learning and more time coding with practical eBooks and Videos from over 4,000 industry professionals

- Improve your learning with Skill Plans built especially for you

- Get a free eBook or video every month

- Fully searchable for easy access to vital information

- Copy and paste, print, and bookmark content

Did you know that Packt offers eBook versions of every book published, with PDF and ePub files available? You can upgrade to the eBook version at packtpub.com and as a print book customer, you are entitled to a discount on the eBook copy. Get in touch with us at customercare@packtpub.com for more details.

At www.packtpub.com, you can also read a collection of free technical articles, sign up for a range of free newsletters, and receive exclusive discounts and offers on Packt books and eBooks.

Other Books You May Enjoy

If you enjoyed this book, you may be interested in these other books by Packt:

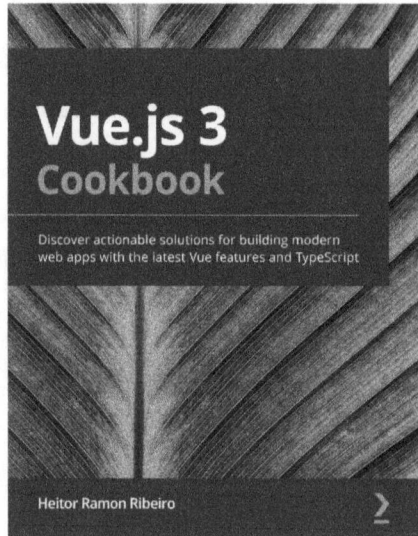

Vue.js 3 Cookbook.

Heitor Ramon Ribeiro

ISBN: 978-1-83882-622-2

- Design and develop large-scale web applications using Vue.js 3's latest features
- Create impressive UI layouts and pages using Vuetify, Buefy, and Ant Design
- Extend your Vue.js applications with dynamic form and custom rules validation
- Add state management, routing, and navigation to your web apps
- Extend Vue.js apps to the server-side with Nuxt.js
- Discover effective techniques to deploy your web applications with Netlify
- Develop web applications, mobile applications, and desktop applications with a single code base using the Quasar framework

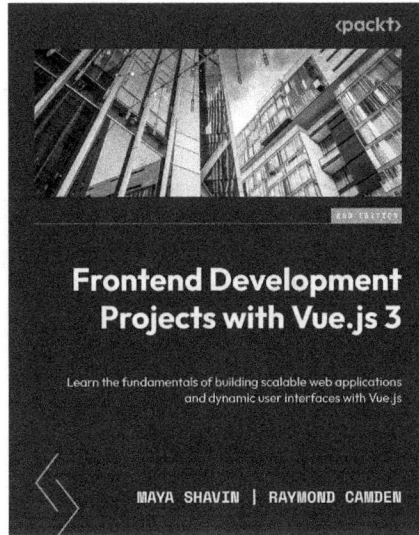

Frontend Development Projects with Vue.js 3

Maya Shavin, Raymond Camden, Clifford Gurney, Hugo Di Francesco

ISBN: 978-1-80323-499-1

- Set up a development environment and start your first Vue.js 3 project
- Modularize a Vue application using component hierarchies
- Use external JavaScript libraries to create animations
- Share state between components and use Pinia for state management
- Work with APIs using Pinia and Axios to fetch remote data
- Validate functionality with unit testing and end-to-end testing
- Get to grips with web app deployment

Packt is searching for authors like you

If you're interested in becoming an author for Packt, please visit `authors.packtpub.com` and apply today. We have worked with thousands of developers and tech professionals, just like you, to help them share their insight with the global tech community. You can make a general application, apply for a specific hot topic that we are recruiting an author for, or submit your own idea.

Share Your Thoughts

Now you've finished *Vue.js 3 for Beginners*, we'd love to hear your thoughts! Scan the QR code below to go straight to the Amazon review page for this book and share your feedback or leave a review on the site that you purchased it from.

https://packt.link/r/1805126776

Your review is important to us and the tech community and will help us make sure we're delivering excellent quality content.

Download a free PDF copy of this book

Thanks for purchasing this book!

Do you like to read on the go but are unable to carry your print books everywhere?

Is your eBook purchase not compatible with the device of your choice?

Don't worry, now with every Packt book you get a DRM-free PDF version of that book at no cost.

Read anywhere, any place, on any device. Search, copy, and paste code from your favorite technical books directly into your application.

The perks don't stop there, you can get exclusive access to discounts, newsletters, and great free content in your inbox daily

Follow these simple steps to get the benefits:

1. Scan the QR code or visit the link below

https://packt.link/free-ebook/9781805126775

2. Submit your proof of purchase
3. That's it! We'll send your free PDF and other benefits to your email directly